ꞯook is to be returned on or before
the    date stamped below

# MRE

Materials Research and Engineering
Edited by B. Ilschner and N.J. Grant

Julian Szekely    Göran Carlsson    Lars Helle

# Ladle Metallurgy

With 137 Figures and 24 Tables

Springer-Verlag
New York  Berlin  Heidelberg
London  Paris  Tokyo

Prof. JULIAN SZEKELY
Department of Materials Science
  and Engineering
Massachusetts Institute of Technology
Cambridge, MA 02139/USA

GÖRAN CARLSSON
MEFOS, The Metallurgical Research Plant
P.O. Box 812
S 951 28 Luleå, Sweden

LARS HELLE
OVAKO STEEL Oy Ab
SF 55100 Imatra, Finland

*Series Editors*

Prof. BERNHARD ILSCHNER
Laboratoire de Métallurgie Mécanique
Département des Matériaux, Ecole Polytechnique Fedérale
CH-1007 Lausanne/Switzerland

Prof. NICHOLAS J. GRANT
Department of Materials Science and Engineering
Massachusetts Institute of Technology
Cambridge, MA 02139/USA

ISBN 0-387-96798-2 Springer-Verlag New York Berlin Heidelberg
ISBN 3-540-96798-2 Springer-Verlag Berlin Heidelberg New York

Library of Congress Cataloging-in-Publication Data
Szekely, Julian, 1934–
    Ladle metallurgy/Julian Szekely, Göran Carlsson, Lars Helle.
      p. cm.—(Materials research and engineering)
    Bibliography: p.
    ISBN 0-387-96798-2
    1. Inoculation (Founding)  2. Steel founding.  I. Carlsson,
  Göran.  II. Helle, Lars.  III. Title  IV. Series.
  TS233.S98 1988
  671.2—dc19                                    88-20017

Printed on acid-free paper.

# Editors' Preface

This book seeks to provide a comprehensive coverage of the important and growing field of ladle metallurgy, including theory, practice, and economics.

During the past decade, major advances have been made in the secondary metallurgy of steel and other metals; indeed, secondary metallurgy, that is, the ladle treatment of molten metals, following the melting and refining steps, has become an important and inevitable part of the overall processing sequence.

Ladle metallurgy is attractive because it can provide an effective means for adjusting and fine-tuning the composition and temperature of the molten products prior to solidification processing. Ladle metallurgy allows us to produce materials of very high purity and will become increasingly an essential process requirement. Indeed, many of the novel casting techniques will mandate steels of much higher cleanliness than those in current practice.

Of course, ladle metallurgy or secondary metallurgy is not limited to steel; indeed, major advances have been made and are being made in the secondary processing of aluminum, aluminum alloys, and many specialty metals.

This book provides, for the first time, a comprehensive treatment of the subject, which includes both the science base and the many practical, real-world considerations that are necessary for the effective design and operation of ladle metallurgy systems. The aim of the theoretical chapter is to provide insight and to develop the fundamental basis of ladle metallurgy systems. The chapter concerning practice reflects the many years of practical experience with the operation of ladle metallurgy systems. Finally, the chapter on economics provides a discussion on both the capital costs and the operating costs of secondary metallurgy systems.

This book should be helpful to students of materials processing and to practicing metallurgists, in both the steel and the specialty metals fields.

Cambridge, MA, USA, July 1988                                   N.J. Grant
Lausanne, Switzerland, July 1988                                 B. Ilschner

# Contents

# 1 Overview of Injection Technology

## Göran Carlsson

## 1.1 Introduction

Treating steel in the ladle is as old as the use of ladles in steelmaking. The main purposes for ladle treatment of hot metal and liquid steel include desulphurization, deoxidation, alloying, and inclusion shape control. One of the first ordinary ladle metallurgical processes was the Perrin process [1], in which steel was tapped into a ladle containing premelted slag. The kinetic energy of the steel was used to produce a large reaction surface and intensive stirring in the ladle.

A very efficient method for the treatment of liquid metal is injection of a powdered reagent or alloy. Already in his time, Sir Henry Bessemer [2] suggested that powdered material should be added to the steel. In the 1930s and 1940s, injection of powdered material was used in many different ways. One example from the iron and steel industry is lime powder injection into hot metal for desulphurization. Petersen and co-workers [3] have reported on the use of burnt lime for desulphurization. In the 1950s, the injection technique was introduced in foundries. The main purpose was desulphurization and alloying with magnesium. During that decade, injection metallurgy was not yet a great success, mainly owing to technical problems. A new era for injection metallurgy started in the late 1960s. The technique was developed and improved in many ways and in many countries, for example, in Sweden, the Federal Republic of Germany, and France, at the same time. The advantages of adding powdered material deeply into liquid metal are that metallurgical operations can thereby be carried out faster, with higher yields, with better reproducibility, and to meet special requirements for the products. In this chapter, some ideas will be given on how injection techniques are used in different metal industries.

## 1.2 Apparatus

An injection equipment system consists schematically of a powder dispenser, transportation hose, and lance (Fig. 1.1). In the industry, it is usually completed with containers and silos for rational handling of the powder. The powder dispenser is a high-pressure vessel with a conical lower part. The powder–gas mixture is pressed through a small hole and into an ejector, where the carrier gas is added. The fluidization of the powder is done in order to break up the powder so that the transportation of it from the dispenser will be uniform. The carrier gas used for

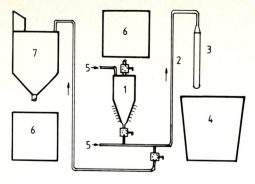

**Fig. 1.1.** Skeleton sketch of an injection installation: 1, powder dispenser; 2, transportation hose; 3, lance; 4, ladle or furnace; 5, carrier gas; 6, powder containers; and 7, buffer container.

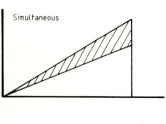

**Fig. 1.2.** Examples of ways to use two dispensers connected to the same hose and lance. Reprinted with permission from ref. 5 (Fig. 1, p. 22:14).

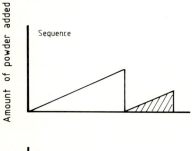

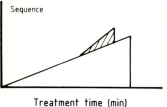

Amount of powder added

Treatment time (min)

steel is argon or nitrogen, while in hot metal nitrogen, air, or in some cases oxygen, is used.

A modern injection station usually contains more than one dispenser, allowing a very flexible treatment [4, 5]. With an injection installation consisting of two dispensers connected to the same hose and lance, either simultaneous or sequence powder additions can be made (Fig. 1.2). The former method can be used for mixing of powders in order to avoid more expensive prepared mixtures and to enable flexible mixing, and for minimizing the vaporization of elements with a high vapour pressure at steelmaking temperatures. Sequence injection can, on the other hand,

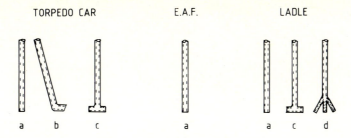

**Fig. 1.3.** Different kinds of lance designs: (a) straight lance, (b) hockey-stick lance, (c) T-hole lance, and (d) 3-hole lance; note that E.A.F. stands for electric arc furnace.

be used for refining, starting with one powder and finishing with another that has a higher affinity to the undesirable element; alloying, made directly after refining without raising the lance (in this way, the precision of the addition will be high); and adding poisonous elements with the lance immersed into the steel. It has been proven that desulphurization can be carried out in two steps with two reagents, giving the same degree of desulphurization compared with refining, which uses only the more expensive reagent, at a cost saving of 25–50% [5].

The injection lance can be constructed in two ways:

1. For the monolithic lance, the ceramic material is cast onto a steel tube.
2. For the sleeve lance, refractory sleeves are piled onto a steel tube.

The outlet of a lance can be constructed in many different ways, according to the type of vessel in which the treatment is carried out (Fig. 1.3).

In order to ensure that a lance will stand the treatment, the following points must be taken into consideration:

1. There is no nozzle blockage.
2. The refractory material has a high thermal shock resistance.
3. The refractory material can withstand high mechanical strain.
4. Slag erosion on the refractory material is controlled.

Today's problems with high costs for lances might be solved in the future by injection through a tuyere in the ladle wall or bottom. Schnurrenberger et al. [6] have developed a system in which a slide gate is used for injection of lime-fluorspar or CaSi (Fig. 1.4). Results have shown that for the same degree of desulphurization the specific quantity of calcium consumed was 0.2 kg less per tonne of steel when using a slide gate than that required for lance treatment. Benefits of using the ladle slide gate nozzle are largely seen as lowered costs for consumption of refractory lining and injected material.

Other systems of tuyeres, slide gates, etc., have been developed in the United Kingdom, the United States, Sweden, France, and the USSR. It looks as if there will be a break-through for injection through the ladle wall or bottom in the near future, but this will of course depend on the economics involved in substituting a

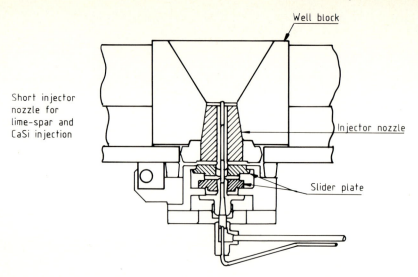

**Fig. 1.4.** Slide gate for injection. Reprinted with permission from ref. 6 (Fig. 3, p. 30:14).

complex ladle tuyere arrangement for a simple lance with high specific refractory costs.

## 1.3 Hot Metal Pretreatment

### 1.3.1 Desulphurization

Treatment of hot metal in a ladle or a torpedo car is done today mostly for the purpose of desulphurization. This is due to the fact that the sulphur content of the steel has an important bearing on the surface quality and mechanical properties of the end product. Therefore, in many integrated steel mills, even for plain carbon steel production, hot metal with no more than 0.020% sulphur is charged into the steelmaking converter. Apart from a qualitative improvement in the steel product, external hot metal desulphurization may constitute a relief for the melting shop permitting a more favorable mode of operation with less basic slag. In the interests of the blast furnace, this allows the use of fuels (coke, coal powder, and oil) with higher sulphur content and the achievement of a lower silicon content in the hot metal, which permits the coke rate to be reduced and the performance to be increased.

Desulphurization reagents can, for economical reasons, be limited to lime, calcium carbide, soda ash, or magnesium. The latter could be salt-coated magnesium granules, in which the salt gives a damping of magnesium evaporation. All of these reagents can be injected into the hot metal. Consumptions of different desulphurization reagents are listed in Table 1.1.

The effectiveness of the reaction between the desulphurization reagent and the

**Table 1.1.** Consumption of desulphurization reagent; sulphur change from 0.050% to 0.015%

| Reagent | Consumption (kg/tonne) | Vessel |
|---|---|---|
| Salt-coated Mg granules | 0.50 | Torpedo |
| CaD 60/40[a] | 5–6 | Torpedo |
| Lime | 7 | Torpedo |
| Lime plus 5% Al | 4–5 | Ladle |
| Lime plus limestone | 5–6 | Torpedo |
| Soda ash | 6 | Ladle |

[a] CaD 60/40: 60% technical calcium carbide and 40% gas former.

sulphur depends on the sulphur concentration. This becomes clear when the sulphur content is under 0.020% and is independent of temperature and the desulphurization reagent.

Lime–limestone mixtures, for example, 57% CaO, 35% $CaCO_3$, 5% C, and 3% $CaF_2$, have been shown to be effective regarding both desulphurization and cost comparison. At Kawasaki Steel [7], desulphurization of hot metal is carried out in torpedo cars of 260–340 tonne using the above mentioned mixture. With an addition of about 5 kg/tonne, the sulphur content is decreased from 0.040% to 0.015%. Compared with $CaC_2$, the cost index for the same desulphurization degree is 1.00 for the lime–limestone mixture and 3.23 for $CaC_2$.

## 1.3.2 Simultaneous Dephosphorization and Desulphurization Combined with Desiliconization

In the 1980s, a growing concern for the dephosphorization process has arisen, mainly because of the degradation of raw materials with respect to phosphorus and the increasing importance of phosphorus control in the steel. Under these circumstances, the necessity for the dephosphorization of hot metal has been recognized particularly by the studies on external dephosphorization by soda ash and lime-based fluxes.

One of the most interesting procedures for external treatment of hot metal is the simultaneous removal of phosphorus and sulphur. This is possible even though a reducing atmosphere is needed for the desulphurization and an oxidixing atmosphere for the dephosphorization, as shown in Fig. 1.5. Simultaneous desulphurization and dephosphorization can be done with the injection of soda ash or lime-based fluxes.

In order to reach a low phosphorus content in the hot metal, the silicon content must be below 0.15% before dephosphorization. The desiliconization is carried out with iron oxide or gaseous oxygen. The efficiency of different desiliconization practices varies between 50–95% depending on how well the hot metal is being stirred and on the reagent being added. Techniques used are overlaying in the blast furnace runner or injection of the reagent into the ladle or torpedo car.

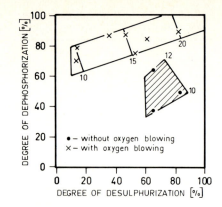

**Fig. 1.5.** Influence of added amount of oxygen on the degree of refining: ☐ Total oxygen addition is 3–4 kg/tonne; ▨ total oxygen addition is 7–10 kg/tonne. Number refer to added amount of lime in kg/tonne. Reprinted with permission from ref. 8 (Fig. 8, p. 476).

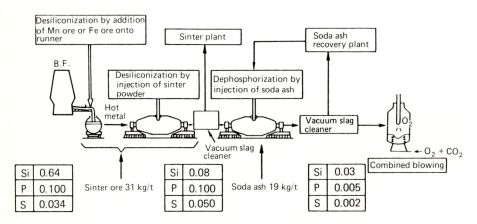

**Fig. 1.6.** The SARP of Kashima Steel Works, Sumitomo Metal [9].

Sumitomo Metal has developed a process for hot metal pretreatment called SARP (Sumitomo alkali refining process) [9], which has been in operation at its Kashima Steel Works since May 1982. The process flow chart is illustrated in Fig. 1.6. It can be seen that, in this process, dephosphorization and desulphurization are simultaneously carried out from $\underline{P} = 0.10\%$ and $\underline{S} = 0.050\%$ before treatment to $\underline{P} = 0.005\%$ and $\underline{S} = 0.002\%$ after treatment by means by injecting (after slag removal) soda ash at 19 kg/tonne hot metal. (Note that underscores indicate that the element is dissolved in the hot metal.) The hot metal has been desiliconized down to $\underline{Si} < 0.10\%$ at the runner and in the torpedo car. The soda ash is recycled through the soda ash recovery plant.

The features of this process, other than the simultaneous desulphurization and dephosphorization, are the minimal manganese loss, the high dephosphorization

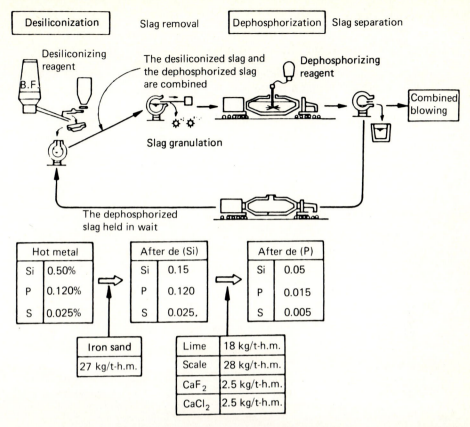

| Desiliconization | Slag removal | Dephosphorization | Slag separation |

**Fig. 1.7.** Process flow of ORP at Kimitsu Works, Nippon Steel [9]; note that "h.m." stands for *hot metal*.

power with a phosphorus distribution ratio of $(P_2O_5)/\underline{P}$ of approximately 1,500 at a slag basicity of $(Na_2O)/(SiO_2) = 3$, and the low decarburization.

In Fig. 1.7, a quicklime flux method called ORP (optimizing the refining process) [9], which has been in operation at the Kimitsu Works of Nippon Steel since September 1982, is illustrated. In this process, the hot metal is desiliconized at the runner to $\underline{Si} < 0.15\%$ and removed of slag, then dephosphorized and desulphurized down to $\underline{P} = 0.015\%$ and $\underline{S} = 0.005\%$ from $\underline{P} = 0.120\%$ and $\underline{S} = 0.025\%$ by injecting 51-kg lime-based flux per tonne of hot metal. Features of this process are, first, the adaptability to all general-purpose steels, not limited specifically to the low phosphorus steels, and second, the easy slag disposal, in which the dephosphorization slag is combined with the desiliconization slag and the mixed slag is subjected to granulation.

Trials at MEFOS (The Foundation for Metallurgical Research, Luleå, Sweden) have shown that a high freeboard is a must for dephosphorization because of the fact that the slag foams so easily. For treatment of 80–100 tonne of hot metal, a freeboard of approximately 1.5 m is needed or a possibility for continuous slag tapping must be constructed.

## 1.4 Steel Refining

### 1.4.1 Deoxidation and Desulphurization

Traditionally, ladle treatments have been classified according to the available equipment. There are vacuum treatments like stream degassing, ladle degassing, RH (Rheinstahl–Hattingen), and DH (Dortmund–Hörde). Then there are arrangements with a heating unit, that is, ladle furnace. In these methods, the whole treatment clearly consists of several successive steps that can be characterized as unit operations. The terms unit process and unit operation are fairly clear in many chemical processes but in metallurgy, for example, in ladle metallurgy, this kind of classification is more complicated. A proposal has been presented by Hólappa [10] and is shown in Table 1.2.

The injection technique can be used when it is suitable to make the addition in powder form, for example, in deoxidation and desulphurization and during alloying. Powders for deoxidation and desulphurization that are industrially used today are listed in Table 1.3. The utilization of the injected elements is extremely efficient as a result of the transitoric phase contact between the reagent and the liquid steel. This is illustrated in Fig. 1.8.

The position and immersion depth of the lance must be correct. Water model experiments with different lance positions have been carried out at MEFOS. It was

**Table 1.2.** Unit processes and unit operations in ladle metallurgy[a]

| Unit process | Unit operation | Reaction mechanisms |
|---|---|---|
| Degassing (H, N) | Vacuum treatment | Vaporization |
| Deoxidation | Top slag treatment | Reactions in reduced pressure |
| Desulphurization | Powder injection | Precipitation reactions |
| Decarburization | Treatment with oxidizing slags | Exchange reactions between metal and slag |
| Dephosphorization | Alloying | Dissolution |
| Composition adjustment | Heating, cooling | Homogenization |
| Temperature control | | Heat transfer |

[a] Reprinted with permission from ref. 10 (Tab. 1, p. 1:4).

**Table 1.3.** Desulphurization agents for powder injection

| Desulphurizing agent | Components | Injected amount (kg/tonne) | Amount (kg/tonne) of agent for $\Delta S/S_0 = 0.90$ |
|---|---|---|---|
| CaSi | 30% Ca, 62% Si, 0.8% Al | 2–4.5 | 3.8 |
| Calcium carbide | 80% $CaC_2$, 15% CaO(50% Ca) | 1–3 | 2.3 |
| Mg(CaO/$CaF_2$) | 5–20% Mg | 1–3 | 2.2 |
| CaO/$CaF_2$ | 90% CaO, 10% $CaF_2$ | 3–6 | 3.0 |
| CaO–$Al_2O_3$ | 50 %CaO, 50% $Al_2O_3$ | 1–5 | 4.0 |
| CaO–$Al_2O_3$–$CaF_2$ | 70% CaO, 20% $Al_2O_3$, 10% $CaF_2$ | 1–5 | 3.0 |

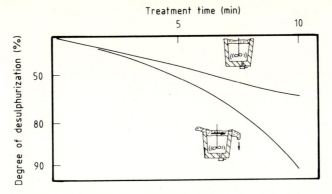

**Fig. 1.8.** Prediction of desulfurization kinetics for different metal or slag contacts [11].

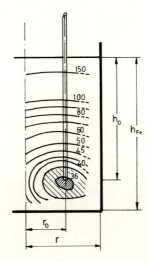

**Fig. 1.9.** Optimal lance position. Figures given are the mixing time in seconds at fixed gas flow rate [11].

found that the shortest mixing time was obtained when the lance outlets were immersed at half the radius of the ladle to a depth of 85–90% of the height of the steel bath. Figure 1.9 shows the optimal lance position.

The importance of immersion depth is shown in Fig. 1.10, which contains results from pilot-plant trials. As evidenced by this figure, the desulphurization results with shallow lance immersion are clearly inferior to those with deep immersion of the lance.

In the case of desulphurization of steel, the mixing power density has proved its importance. Figure 1.11 illustrates how the speed of sulphur removal, represented by the rate constant, depends on the power density, here represented by the gas flow rate per tonne of steel. When applying powder injection, the amount of carrier gas needed results in power densities to the right of the knee of the curve, while ordinary top slag addition and gas purging result in power densities to the left of the knee.

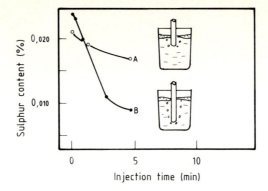

**Fig. 1.10.** Desulfurization results using different injection techniques: (a) shallow injection, no gas purging; and (b) deep lance injection. Reprinted with permission from ref. 12 (Fig. 1, p. 8).

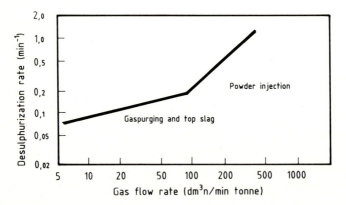

**Fig. 1.11.** Desulfurization rate constant, $K_s$, as a function of the gas flow rate [11].

When performing top slag addition and gas purging with very high gas flow rates, the rate of desulphurization equals that of powder injection.

The desulphurization, or control of the sulphur content, can hardly be done with good reproducibility in various vessels without knowing something about the thermodynamics and rate phenomena involved. For the desulphurization of steel, these reactions have to be taken into consideration:

$$\underline{Ca} \quad \text{or} \quad [Ca] + \underline{S} = (CaS) \tag{1.1}$$

$$(CaO) + \underline{S} = (CaS) + \underline{O} \tag{1.2}$$

$$[Mg] + \underline{S} = (MgS) \tag{1.3}$$

$$(O^{2-}) + \underline{S} = (S^{2-}) + \underline{O} \tag{1.4}$$

It is not yet clearly shown that reactions (1.1) and (1.3) actually take place when injecting calcium or magnesium in steel melts at 1600°C. Calculated and analyzed sulphur distributions from samples taken after vigorous argon purging or powder

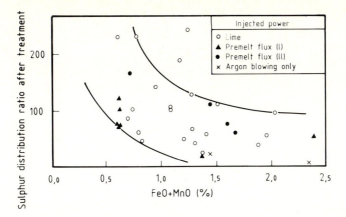

**Fig. 1.12.** Influence of FeO and MnO in slag on the sulfur distribution ratio after treatment. Reprinted with permission from ref. 13 (Fig. 9, p. 21:15).

injection have been compared. For high sulphur distribution values, well-reduced, lime-rich slags are necessary.

One may repeat the old saying "for a good desulphurization, a good slag of high basicity is needed." The detrimental effect of easily reduced oxides such as FeO and MnO is demonstrated in Fig. 1.12. The tapping operation of the furnace has to be specially reviewed to ensure a successful ladle treatment.

With CaSiMg injection, desulphurization is reported to proceed very quickly at the early stages of injection [14]. After about 200 g Mg per tonne, the desulphurization is claimed to stop at a certain level, and continued injection does not promote sulphur removal. When comparing CaSi injection with CaSiMg injection, Johansson [14] has reported that in the case of CaSi injection the time needed to reach equilibrium is about 3 min or more, depending on the degree of desulphurization, whereas it reduces to about 2 min when calcium is partially replaced by magnesium.

The initial rate of desulphurization due to CaO + Mg injection is reported to be improved by a factor of about two as compared to that by CaSi injection or Ar stirring with top slag. The increased usage of Mg in the mix is, however, limited because of the risk of increased splashing of metal. The sulphur levels achievable as a function of time with the injection of the CaO + Mg mix, CaSi, and $CaCN_2$ are presented in Fig. 1.13. One advantage of CaSi, if compared with the CaO + Mg mix and $CaCN_2$, is its better flowability.

It is important to remember that, for example, for the aim of antihydrogen-induced-cracking steel production, it is necessary to control even the Ca/S ratio at an appropriate level in addition to controlling the sulphur content depending on the steel grade [15].

It has been the experience of Carlsson and Helle [15] that an injection of CaO + $CaF_2$ type slag mixture ensures the best degree of deoxidation and cleanliness from oxide inclusions in the steel. During the experiments they carried out [15], the total oxygen content as well as the number of different inclusion sizes

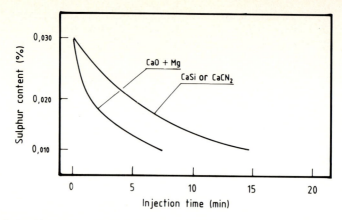

**Fig. 1.13.** Relationship between desulfurization with CaO + Mg, CaSi, and CaCN$_2$. Reprinted with permission from ref. 14 (Fig. 9, p. 18:25).

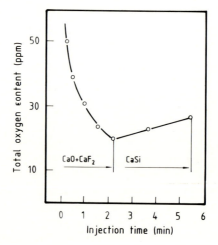

**Fig. 1.14.** Change in total oxygen content of the steel during a typical multicomponent injection. Reprinted with permission from ref. 15 (Fig. 5, p. 21).

decreased greatly during CaO + CaF$_2$ injection. Surprisingly, the opposite was true for CaSi; it resulted in an increase in the total oxygen content by an average value of 2 ppm per kilogram of CaSi injected per tonne of steel (see Fig. 1.14). As a consequence, injection of CaSi has been reported to increase the number of oxide inclusions, particularly when the oxygen content of the steel is already low at the beginning of the injection. There is generally an increase in the inclusion size.

However, no modification of alumina to calcium aluminates was achieved as a result of the CaO + CaF$_2$ injection, and a few sulphide inclusions were still found after the injection.

Injection of CaSi has the unique advantage that it leads to the modification of manganese sulphide inclusions to globular calcium sulphide and oxy-sulphides as well as that the aluminates are modified into globular calcium aluminates, which improve the castability. The minimum amount of CaSi required to be injected for attaining this objective has been mentioned as 0.9 kg/tonne. It has to be understood that the necessary amount depends on the prevailing metallurgical conditions in each particular case. More of this will be discussed under the next heading. The inclusion modification phenomena will be elaborated in Chapter 3.

Observations by Moriya et al. [16], Helle [15], and Johansson [14] show that the degree of desulphurization achieved by injection could be similar for $CaO-CaF_2$ (with or without $Al_2O_3$) flux, CaSi, a combination of these two, $CaC_2$, $CaCN_2$, or $CaO + Mg$ mix. The rate of desulphurization is reported to be the lowest for $CaC_2$, and then becomes higher for premelted $CaO-CaF_2-Al_2O_3$ flux, and then is highest for CaSi injection.

One of the problem areas of injection of Ca-bearing materials is the pickup of hydrogen and nitrogen by the steel. The amount of hydrogen pickup is proportional to the amount of desulphurization agent injected (Fig. 1.15).

In contrast to this data, which show that the pickup is greater for $CaC_2$ and CaSi injection, the data in Table 1.4 show that it is higher for CaO-bearing flux.

Injection of CaSi or CaSi-containing mixtures is usually associated with more nitrogen pickup because of the stirring effect at the metal–atmosphere interface caused by the evaporation of calcium. The degree of pickup could be 10–15 ppm, depending on the specific quantity of powder injected [18].

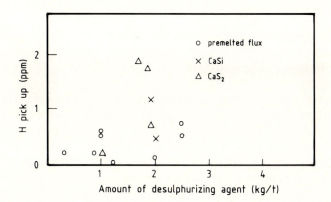

**Fig. 1.15.** Relation between amount of desulphurizing agent and hydrogen pickup. Reprinted with permission from ref. 17 (Fig. 9, p. 16:22).

**Table 1.4.** Hydrogen results[a]

| Type of treatment | Number of heats | Hydrogen (ppm) | | Pickup (ppm) |
|---|---|---|---|---|
| | | Before injection | After injection | |
| 85% CaO + 15% $CaF_2$ | 6 | 2.5 | 3.7 | +1.2 |
| 100% CaSi | 5 | 3.2 | 3.9 | +0.7 |

[a] Reprinted with permission from ref. 20 (Tab. 10, p. 41).

**Table 1.5.** Average aluminium loss[a]

| Powder injected | Al loss (%) | Numbers of heats |
|---|---|---|
| 100% CaSi | $0.014 \pm 0.014$ | 50 |
| 85% CaO + 15% CaF$_2$ | $0.032 \pm 0.034$ | 14 |

[a] From Tivelius et al. [20].

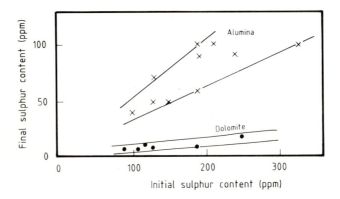

**Fig. 1.16.** Influence of ladle refractory material on desulphurization efficiency [11].

The average aluminium loss during ladle treatment is the smallest when injecting CaSi as compared with Ar stirring or injection of CaO–CaF$_2$. The results for aluminium loss in a dolomitic ladle are shown in Table 1.5. Aluminium loss is reported to be higher in an acid-lined ladle because of leakage of oxygen from the silica lining [19].

Based on the relative merits and drawbacks of the various reagents and ladle treatment processes discussed so far, it appears that for the purposes of desulphurization, deoxidation, and inclusion control in many grades of clean steels, it may be adequate to resort to either gas injection with a CaO–CaF$_2$–Al$_2$O$_3$ or CaO–SiO$_2$–Al$_2$O$_3$ top slag or injection of a CaO-based powder followed by injection of an appropriate amount of CaSi, unless the steel grade is susceptible to hydrogen effects. Though experiences at various places have revealed that a similar and high degree of desulphurization has been achieved through gas injection or powder injection with various fluxes, it is felt that more attention may be paid in this area in the future to optimize the flux composition for steel refining efficiencies.

The ladle refractory has been found to have a great influence on the desulphurization result. This is due to the fact that, during desulphurization, the oxygen content of the steel is lowered, and consequently, oxides in the lining material that can be reduced by Ca will be so and reoxidation of the melt will occur. A comparison of desulphurization results in ladles lined with different refractory materials is shown in Fig. 1.16. The difference in desulphurization degree obtained because of the refractory material in the ladle is clear.

During injection treatment, the temperature of the steel decreases. The measured temperature drop is normally on the order of $1-3°C/min$ for 50-tonne ladles and $10-15°C/min$ for 7-tonne ladles. Apart from the scale, the heat losses depend on the operation performed in the ladle, the steel grade treated, etc. The following practical means can be used to stabilize and minimize heat losses:

1. sufficient insulation of ladle walls,
2. efficient preheating and monitoring of heat input,
3. use of ladle cover during treatment, and
4. short ladle cycle time.

In-ladle injection techniques have been carried out successfully in recent years and provide reliable reproducible desulphurization rates with low oxygen contents, with the use of calcium. The prerequisites, as regards steel, for high desulphurization are:

1. a low oxygen and high aluminium content in the steel,
2. plunging of the lance to a large depth (see Fig. 1.9 and 1.10).
3. a low silicon content in the lining of the ladle (see Fig. 1.16), and
4. a low FeO content in the ladle slag, with a high basicity index (see Fig. 1.12).

### 1.4.2 Inclusion Shape Control

Once the sulphur is precipitated from the liquid steel during solidification, it will exist in two principal forms. The first one of these is a spheroidal inclusion, a more or less harmless phenomenon resulting from low plasticity during rolling. The other is intergranular films—or platelets—which are plastic during hot rolling.

When an element with a high affinity for sulphur, such as calcium or zirconium (CaSi, ZrSi), is introduced by injection, this element will react with sulphur-forming sulphides. Thereby, for example, spheroidal complex calcium sulphides can be formed instead of manganese sulphides, which would deform on rolling. The purer the MnS, with respect to calcium, the more plastic it becomes as evidenced by the increase of the sulphide shape factor ($L/W$) with higher content of pure MnS.

Hence, the injection of CaSi may result in complete modification of MnS inclusions [20]. Some practical results of desulphurization with CaSi are given in Refs. 21–24 and in Fig. 1.17. From this figure, the zone can be defined where, after treatment, one observes in the liquid steel samples a complete transformation of sulphide inclusions.

When employing metering tundish nozzles, the steel cannot be killed with aluminium because of the formation of alumina slag inclusions and the subsequent clogging at the tundish nozzles. This is caused by attachment of the alumina inclusions to the nozzle wall and high-temperature sintering of alumina inclusions to each other. For this reason, the maximum total aluminium content is limited to around 0.004%, and consequently, the steel is killed normally with Si–Mn to avoid the clogging problem.

For high grade steels, the degree of deoxidation with Si–Mn is not always acceptable, as has been experienced so many times at various works when defects like seams, torn corners, or poor surface finish appear. Thus, there is a strong need for

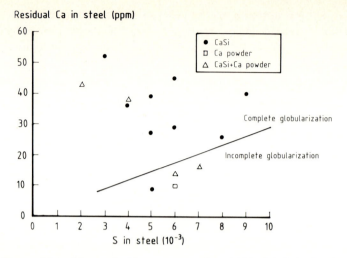

**Fig. 1.17.** Influence of the calcium and sulfur contents on the sulphide globularization. Reprinted with permission from ref. 21 (Fig. 10).

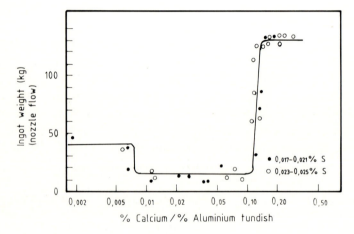

**Fig. 1.18.** Effect of calcium on the flow of aluminum deoxidized steel through nozzles. Reprinted with permission from ref. 25 (Fig. 1, p. 14).

a better deoxidation with aluminium. This is particularly true for qualities such as high carbon wire rod, boron steel, cold heading steel, and low silicon, low carbon steel.

Since calcium has a higher affinity for oxygen than aluminium, the dendritic alumina inclusions and also the Al–Mn silicates are partially or completely eliminated when the steel is subjected to CaSi injection. Instead, inclusions of calcium aluminate or calcium oxide are formed that are spheroidal in shape and do not form networks or attach to the nozzle wall or to each other.

One of the important factors when aiming for alumina modification is the rela-

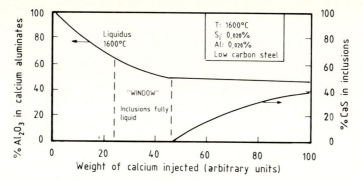

**Fig. 1.19.** Schematic diagram showing the changes in the composition of inclusions during injection of calcium. Reprinted with permission from ref. 26 (p. 17).

**Table 1.6.** Number of worldwide powder injection units [38]

| Location | Percent of basic oxygen furnace shops with powder injection | Percent of electric arc furnace shops with powder injection |
|---|---|---|
| Europe | 51 | 27 |
| North America | 27 | 52 |
| Japan | 44 | 8 |
| Others | 30 | 12 |

tionship between calcium and aluminium in the steel. With small additions of calcium, nozzle clogging is actually increased but as the calcium addition is sufficient the blockage is eliminated [25]. This phenomenon is clearly demonstrated in Fig. 1.18. From this figure, it can be concluded that the Ca/Al ratio must be at least 0.15 to get the full benefit of calcium.

Davies and co-workers [26] have reported that excess additions of calcium could form calcium sulphide inclusions, which are solid at steelmaking temperatures (1500–1600°C) and will cause tundish nozzle clogging. Therefore, there is a safe "window" of calcium additions (Fig. 1.19) below which the low melting point aluminates suited to continuous casting will not be formed and above which calcium sulphide is produced. This phenomena will be elaborated further in Chapter 3.

### 1.4.2.4 Statistics

The number of powder injection units worldwide, according to Nakayama [38], are given in Table 1.6. The expected worldwide growth rate from 1985 to 1987 was

$$\begin{array}{ll} \text{BOF shops} & 41\% \text{ to } 50\% \\ \text{EAF shops} & 18\% \text{ to } 23\% \end{array}$$

There are several reasons for the larger growth rate of injection in basic oxygen

**Table 1.7.** Inclusion parameters relevant for steel properties

| Steel property | Main influence by | | Most significant inclusion parameters | | | | | | |
|---|---|---|---|---|---|---|---|---|---|
| | Oxides | Sulphides | Wt %O | Wt %S | Aₐ oxides | Aₐ sulphides | N₂(D) | P | F |
| Machinability | X | X | X | X | X | X | X | X | X |
| Polishability | X | | X | | X | X | X | | |
| Corrosion | | | | | | | | | |
| Pitting | (X) Duplex | X | | X | (X) Duplex | X | (X) | | |
| Stress corrosion cracking | (X) | X | | X | | X | X | | |
| Short transverse ductility | (X) | X | | (X) | | X | | X | X |
| Lamellar tearing | | X | | | | | | X | |
| Fracture toughness | X | X | | | | | X | X | |
| Hot ductility | (X) | X | (X) | (X) | (X) | (X) | X | X | |
| Fatigue | X | | X | | | | (X) | X | |
| Thermal fatigue | (X) | X | | | | | X | | |
| Welding | X | X | X | X | X | | X | | X |
| Cleanliness | X | X | X | X | X | X | X | | |

furnace (BOF) shops. Powder injection can be a relatively inexpensive way to go for production increase and higher steel grades, such as pipeline steels. The consumption costs for the lance, for example, will be very low if the heat size is over 200 tonne [39]. Also, the need for powder injection is stronger in flat products, such as linepipe, which are mainly produced by BOF.

### 1.4.3 Mechanical Properties and Machinability

The influence of nonmetallic inclusions on mechanical properties of steel has been widely studied. A very good overview is given by Nicholson and co-workers [22]. Tables 1.7 and 1.8 summarize relevant information for metallurgists.

The progress made in reducing machining time in mechanical engineering steel, especially for the motor industry, is partly attributable to the addition of elements such as S, Pb, Ca, Se, and Te, of which the main effect of the last three is to modify the morphology of the sulphides. To enhance the role of sulphur in cutting-tool protection, attempts have been made to use oxide inclusions formed in the course of deoxidation, particularly combined with calcium to form globular inclusions with a calcium aluminate core surrounded by a calcium sulphide peel.

Extensive tests carried out by Ovako Steet Oy Ab [27] since 1976 have shown a considerable improvement in the life of sintered carbide tools when machining steel containing the above mentioned inclusions (see Fig. 1.20). In particular, reduction in the flank wear has been noticed (Fig. 1.21), which is often a decisive factor of the machinability. The beneficial effect of calcium aluminates on the sintered carbide machinability has been verified at the cutting speed range of 100 to 300 m/min,

**Table 1.8.** Examples of ladle treatment relevant to steel properties[a]

| Property | Metal | Refining agents | References |
|---|---|---|---|
| Hot workability | High alloy | Ca + REM | Met. Trans. B(1982):13B, p. 603 |
| Hydrogen-induced cracking (HIC) | Si–Al killed high Mn | Flux/Ca | SCANINJECT III (Moriya), SCANINJECT III (Iida) |
| Toughness HIC | gas resistant line pipe | Ca | 66th Steel Making Conference (Murakami) |
| Impact strength | 10MoNiCuSiREM | Ca + REM | SCANINJECT III (Naizao) |
| Machining | | Ca | SCANINJECT III (Backman) |
| Ductility | EW pipes | Ca | SCANINJECT III (Ototani) |
| Magnetism | Si steel | Flux + REM | SCANINJECT III (Iida) |
| Corrosion | EW pipes | Ca | SCANINJECT III (Ototani) |
| General | Offshore steel | | SCANINJECT III (Grip) |

[a] For a more complete search, use available data, for example, Met. Abstract.

although in some production operations, improved machinability has also been observed at lower cutting speeds.

At present, the possibilities offered by calcium deoxidation are a field of active research mainly directed toward:

1. Steels for high-speed machining, with or without grain control, especially for the motor industry (rack gears for steering columns, universal joints, sundry shafts,

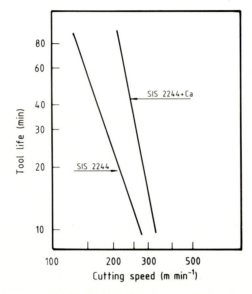

**Fig. 1.20.** The $v$–$T$ curves for normal SS 2244 type steel containing calcium aluminate based inclusions. Carbide tool P10, work piece hardness 250 HB [27].

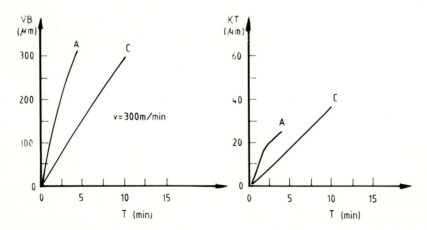

**Fig. 1.21.** The effect of calcium aluminate (C) and anorthite (A) inclusions on the flank and crater wear of a carbide tool (P10) in turning. Work piece materials SS 2244 (250 HB). $VB$ = flank wear, $KT$ = crater wear [27].

etc.). This field is an immediate and economically attractive application as regards machining of rolled or forged bars for mechanical applications (railway axles, etc.).

2. Steels for low-speed machining (milling, cutting, drilling) of gear wheels in particular, normally with controlled grain, either by aluminium or by Nb or V. Recent research has shown the possibility of making steels with plastically machinable inclusions, not only for high-speed but also for low-speed machining. Generally speaking, the in-ladle injection technique seems particularly promising in the production of these steels.

### 1.4.4 Dephosphorization

During recent years, dephosphorization of the steel, after the converter or EAF, has been introduced in order to meet higher demands on the mechanical properties of the steel. Two different refining operations are used depending on the steel grade produced:

High Cr steel:   Refining under reducing conditions with Ca
Low alloy steel:   Oxidizing refining

### 1.4.4.1 Refining with $CaC_2$

Experiments carried out (e.g., with 6–7 tonne of liquid high Cr steel and injection of the reagent see Fig. 1.22) have shown that a refining degree of 40–50% is possible [28].

The amount of $CaC_2$ added should exceed 20 kg/tonne and be preferably around 25 kg/tonne. This is because of the fact that calcium has a higher affinity to oxygen, sulphur, arsenic, antimony, tin, and nitrogen than to phosphorus and, consequently, reacts first with these elements.

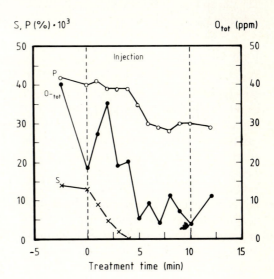

**Fig. 1.22.** Typical variation in chemical composition of the melt [28]: $m_{steel} = 6.0$ tonne, $m_{CaC_2} = 30.4$ tonne, $T_{initial} = 1641°C$, $C_{initial} = 1.02\%$

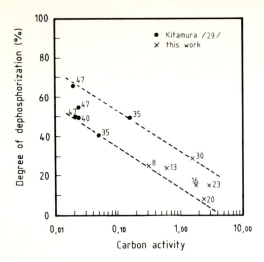

**Fig. 1.23.** Influence of carbon activity upon the degree of dephosphorization. The numbers refer to the amount of $CaC_2$ added [28].

   Carbon activity has been found to influence the results greatly (see Fig. 1.23). An activity below 0.1 is a must in order to reach a high degree of dephosphorization.

   The top slag must be separated from the melt as soon as possible after finishing the dephosphorization. The slag must then be oxidized in order to avoid phosphine ($PH_3$) formation at room temperature. Phosphine is an extremely poisonous gas. Additions of oxygen or iron ore are two effective ways.

### 1.4.4.2 Oxidizing Refining

Daido Steel Co. Ltd. [30] has developed a refining operation called ELVAC (electric furnace, ladle furnace, vacuum and caster) in order to reduce the impurity levels in ball-bearing steels. Immediately after tapping, dephosphorization is carried out. The reagent used is a blend of $CaO-CaF_2-FeO$, which is injected into the liquid steel. The phosphorus level is decreased from 60 to 20 ppm. The injection is followed by a deslagging operation.

### 1.4.5 Alloying

One important motive for the installation of the ladle metallurgy system is the possibility of relieving the primary furnace from all time-consuming tasks. For instance, all alloying operations can be moved to the ladle. Most elements can be added on top of the steel bath during the desulphurization injection with recoveries close to 100%. Because of the large specific stirring effect from the carrier gas, an effective homogenization takes place in a short time. However, a few microalloying elements, often with high affinity to oxygen, can be injected with excellent results. To this group belong FeB, FeTi, FeV, FeNb, S, Te, and Pb [31]. This alloying can be carried out from an extra dispenser in a multidispenser system.

   Trim alloying, that is, achievement of an exact chemistry, can be carried out with extremely high accuracy with powder injection. This procedure enables savings of alloys, a minimum of heats out of specification, and a possibility to meet demanding customer requirements. The trim alloying normally takes place during the last minutes of the injection procedure in order to save process time.

## 1.5 Foundry

As already mentioned, the injection technique was used for desulphurization and magnesium alloying in foundries in the 1950s, but the technique failed because of technical problems. Today, a new era has started in foundries, especially in steel foundries.

Special advantages exist for injection treatment of steel in foundries:

1. The low amount of nonmetallic inclusions and the low oxygen activity and sulphur content improve the castability. The riser length increases and the amount of hot tears decreases.
2. By injection of CaSi, the ductility of the steel will increase (elongation, impact value), but the tensile strength may decrease.

Certain considerations have to be made when injecting powder into small ladles [32]:

1. The ladle should preferably be preheated or in continuous circulation. A cold ladle will absorb too much heat during the agitation caused by the injection. On the other hand, if the ladles are preheated properly, the heat losses are predictable. The temperature drop caused by the injection treatment has to be compensated by an overshot of temperature in the primary furnace.
2. A smooth powder/gas flow is a must. Excessive gas contributes to an accelerated temperature drop and causes splashing. In this respect, multinozzle lances resting on the bottom of the ladle will contribute to a calm injection.

An increasing number of foundries worldwide are using injection techniques to meet higher quality standards for their cast steel. Injection techniques can also be used in the cast iron foundries. When the influential parameters, for example, temperature drop, lining material, and melt analysis, are under control, one can expect certain results from various treatments, such as

1. desulphurization—decreasing the sulphur level from 0.10% to 0.01% with an addition of 0.5–0.7% $CaC_2$;
2. alloying of magnesium—a magnesium recovery of 70–80%; and
3. recarburization—a carbon recovery of over 90%.

With the techniques of today, it is possible to handle the temperature drop even for ladle sizes down to 4 tonnes, and with the technological know-how from the steel industry, it will be possible to inject powder material into liquid iron with good results.

## 1.6 Copper

Lehner and co-workers [33] have reported on the use of powder injection for refining of matte and blister copper. Experiments with injection of soda–lime mixtures reduced Sb and As 50% and 90%, respectively. Table 1.9 summarizes the results achieved by different researches.

Different kinds of slag mixtures can be injected in order to remove different ele-

**Table 1.9.** Results from powder injection refining

| Reagent | Element removed | Final content | Reference |
|---------|-----------------|---------------|-----------|
| Soda ash | As | 0.02% | 34 |
|          | Sb | 0.03% |    |
| P–Cu alloy | Pb | 0.01% | 35 |
| $CaC_2$ | O | 0.001–0.0015% | 36 |

**Table 1.10.** Alloying of aluminium

| Addition | Average yield (%) |
|----------|-------------------|
| Si powder | 99.5 |
| Mn powder | 98.0 |
| Mg granules (salt-coated) | 98.4 |

ments from copper. Helle [37] has, in a literature survey, reviewed the metallurgical results that can be achieved. The following slags have been used with success:

1. alkali slags for the removal of arsenic and antimony,
2. silicate slags for the removal of bismuth, antimony, and nickel,
3. borate slags for the removal of nickel and lead, and
4. phosphate slags for the removal of arsenic, antimony, and lead.

When special slags are used, attention must be paid to the wear of the refractory lining.

## 1.7 Aluminium

Over the last 15 years, the aluminium industry has made attempts to adopt injection technology. Trials have been carried out in many countries, for example, in Japan, Norway, and Great Britain. For about 10 years, the injection technique has been used in production.

The injection technique is mostly used for alloying. The yield of the added element is high, as can be seen in Table 1.10.

Injection alloying has shown a lower number of inclusions in the aluminium than when alloys are added in lump form. The salt coating on magnesium granules acts as a refining flux with respect to oxides and carbides.

Experiments with injection of iron, zinc, chromium, and copper powders into liquid aluminium have also been reported.

## 1.8 Ferro Alloy

In recent year, injection technology has been used in the ferro alloy industry. The purposes of the treatment have been alloying or refining.

Silicon oxide is injected for refining to both FeSi and Si metal. Oxygen from silicon oxide is used for the removal of elements with higher oxygen affinity than silicon, for example, aluminium. Nitrogen can be used as a carrier gas, and the lance can be made of graphite. Calcium carbide can be used for desulphurization of FeSi. With an addition of $1-3$ kg $CaC_2$ per $0.01\%$ S and tonne of FeSi, the sulphur content can be decreased from $0.07-0.03\%$ to $0.03-0.006\%$. This is without any unusual metal losses.

# References

1   J. Strauss, "The Perrin process," Proc. 4th Electric Furance Conference, Pittsburgh, Pennsylvania, December 5–7, pp. 236–240 (1946)
2   H. Bessemer, "An autobiography," Offices of Engineering, London (1905)
3   O. Petersen et al., "Entschwefelung von Thomasroheisen," Stahl und Eisen 60, 677–684 (1940)
4   P.J. Koros et al., "The lime-mag process for desulphurization of hot metal," Proc. SCANINJECT I, Luleå, Sweden, June 9–10, pp. D15:1–D15:13 (1977)
5   G. Carlsson, "Refining and alloying of steel by multi-component injection technique," Proc. SCANINJECT III, Luleå, Sweden, June 15–17, pp. 22:1–22:20 (1983)
6   E. Schnurrenberger et al., "Steel treatment in the ladle by injection through a slide gate nozzle," Proc. SCANINJECT IV, Luleå, Sweden, June 11–13, pp. 30:1–30:22 (1986)
7   K. Taoka et al., "Development of hot metal desulphurization with limestone based flux," Proc. 67th Steelmaking Conference, Chicago, April 1–4, pp. 335–339 (1984)
8   G. Carlsson et al., "Dephosphorization of hot metal with lime based fluxes—Experiments in 5 t scale," Proc. Shenyang Symp. on Inj. Met. and Second Ref. of Steel, Shenyang, People's Republic of China, September 19–21, pp. 467–483 (1984)
9   S. Anezaki et al., "Hot metal pre-treatment," Trans. ISIJ 25, pp. 652–658 (1985)
10  L.E.K. Holappa, "Review of ladle metallurgy," Proc. SCANINJECT II, Luleå, Sweden, June 12–13, pp. 1:1–1:24 (1980)
11  G. Carlsson et al., "Entwicklung der Einblasmetallurgie in Skandinavien," Verein Deutscher Eisenhüttenleute, Fachausschussberiecht 2.023, pp. 45–48 (1985)
12  M. Turunen, "Injection for treatment of steel in a ladle," Jernkontorets Annaler 159, pp. 7–12 (in Swedish) (1975)
13  K. Wada et al., "Investigation of desulphurization and deoxidation in injection metallurgy," Proc. SCANINJECT II, Luleå, Sweden, June 12–13, pp. 21:1–21:15 (1980)
14  R. Johansson, "Injection of powdered material in a 120 ton ladle in Smedjebacken," Proc. SCANINJECT I, Luleå, Sweden, pp. 18:1–18:25 (1977)
15  L. W. Helle et al., "Multi component injection—Development of the technology and results of trials at MEFOS," Scand. J. Metallurgy 14, pp. 18–24 (1985)
16  A. Moriya et al., "Steel quality improvement by flux injection," Proc. SCANINJECT III, Luleå, Sweden, June 15–17, pp. 32:1–32:34 (1983)
17  A. Moriya et al., "Ladle treatment by injection," Proc. SCANINJECT II, Luleå, Sweden, June 12–13, pp. 16:1–16:26 (1980)
18  M. Devaux et al., "Control of non-metallic inclusions in steel by shallow silico-calcium injection in the ladle," Proc. SCANINJECT II, Luleå, Sweden, June 12–13, pp. 31:1–31:33 (1980)
19  G. Folmo et al., "Desulphurization of steel in an acid ladle using CaSi and lime-fluorspar mixtures," Proc. SCANINJECT II, Luleå, Sweden, June 12–13, pp. 15:1–15:18 (1980)
20  B. Tivelius et al. "The influence of different desulphurizers on the process parameters and properties of solid material during TN ladle treatment," Iron & Steelmaker, pp. 38–46 (November 1979)
21  C. Marique, "Industrial experience gained in Belgium with various calcium treatments of steel— Metallurgical and economical aspects," Proc. ECE Seminar on the Economical Aspects of Secondary Steelmaking, Dresden, East Germany, June 15–19, paper No. R.5 (1987)
22  A. Nicholson et al., "Non-metallic inclusions and developments in secondary steelmaking," Ironmaking & Steelmaking 13, pp. 53–69 (1986)

23  R. Väinölä et al., "Establishment of calcium treatment practices at Ovako—evaluation of alternative methods," Proc. SCANINJECT IV, Luleå, Sweden, June 11–13, pp. 23:1–23:20 (1986)

24  J.M. Henry et al., "Different types of calcium treatment as a contribution to the development of the continuous casting process," Proc. SCANINJECT IV, Luleå, Sweden, June 11–13, pp. 24:1–24:18 (1986)

25  G.M. Faulring et al., "Steel flow through nozzles—influences of calcium," Iron & Steelmaker, pp. 14–20 (February 1980)

26  I.G. Davies et al., "Secondary steelmaking developments on engineering steels at Stockbridge works," Proc. Conf. Secondary Steelmaking for Product Improvement, London, Great Britain, pp. 17:1–17:13 (1984)

27  V. Ollilainen, "The effect of Ca-treatment on the machinability of steel," Proc. Swedish Symposium on Non-Metallic Inclusions in Steel, Södertälje, Sweden, pp. 429–450 (1981)

28  G. Carlsson, "The dephosphorization of stainless steel with $CaC_2$—Experiments in 6 tonne ladle," Proc. 4th Japan–Nordic Countries Joint Symposium on Science and Technology of Process Metallurgy, Tokyo, Japan, November 17–18, pp. 279–298 (1986)

29  K. Kitamura et al., "Production of low phosphorous stainless steel by the reducing dephosphorization process," Trans. ISIJ 24, pp. 631–638 (1984)

30  T. Yajima et al., "Improvements of plant economy by secondary steelmaking processes applied for speciality steels," Proc. ECE Seminar on the Economic Aspects of Secondary Steelmaking, Dresden, East Germany, June 15–19, paper No. R. 7 (1987)

31  B. Tivelius et al. "Efficient injection metallurgy system—The importance of raw materials selection and equipment design," information paper

32  Scandinavian Lancers AB, "Injection technology—from steel mill to steel foundry," information paper

33  T. Lehner et al., "Injection metallurgy for the refining of matte and blister copper," Proc. 23rd Annual Conference of Metallurgists, Quebec, Canada, August 19–22, lecture No. 2 (1984)

34  S. Lundqvist, "Removal of alkali from blister copper," Boliden Metall AB, Skelleftehamn, Sweden, Technical Report No. 6/55 (in Swedish)

35  J.E. Stolarczyk, "The removal of lead from copper in fire refining," J. Inst. Met. 86, pp. 49–58 (1957–1958)

36  A. Bydalek, "Undersuchungen über die Raffination von Kupfer und Kupferlegierungen durch Kalziumkarbid," Neue Hütte 22, pp. 663–665 (1977)

37  L.W. Helle "Injection metallurgy—copper," Jernkontoret report No. D323, (1980) (in Swedish)

38  M. Nakayama, IISI, information given at the ECE Seminar on the Economic Aspects of Secondary Steelmaking, Dresden, East Germany, June 15–19 (1987)

39  S. Gustafsson, Scandinavian Lancers International AB, private communication

# 2 The Fundamental Aspects of Injection Metallurgy

Julian Szekely

## 2.1 Introduction

In this chapter, we shall examine some of the fundamental aspects of injection metallurgy. While the majority of readers will be more interested in the practical and economic aspects of injection processes, with emphasis on "What can be accomplished?" and "At what cost?", there are sound reasons for examining the theoretical basis of these operations.

One is that certain aspects of these operations may be predicted from first principles, often performing relatively simple calculations or using well-established relationships. Slag–metal equilibria, purge gas requirements, the stirring power needed, and the like, may be cited as representative examples.

Perhaps more important, a fundamentally based understanding provides a great deal more insight into the behavior of these systems and guidance regarding the optimization of existing operations and the development of new process concepts.

Figure 2.1 shows a schematic sketch of injection systems employed in ladle metallurgy. It is seen that the principal components of this operation may be divided into the following:

1. powder handling and delivery systems;
2. powder–melt interactions; and
3. reaction kinetics, slag–metal reactions, and mixing.

In presenting the theoretical framework for the study of these phenomena, we shall consider these main subdivisions:

1. thermodynamic considerations,
2. powder handling and melt–powder interactions, and
3. process kinetics and mixing.

## 2.2 Thermodynamics

Thermodynamics enables us to calculate the composition of phases in equilibrium; thus, on the basis of thermodynamic considerations, we can predict the theoretical requirements for deoxidation, desulfurization, or dephosphorization agents for a given set of circumstances. By the same token, we can calculate the theoretical amount of purge gas that would be needed to reduce the hydrogen or nitrogen content of steel for a given set of conditions.

Some typical situations of relevance are sketched in Fig. 2.2. Three points need emphasis here.

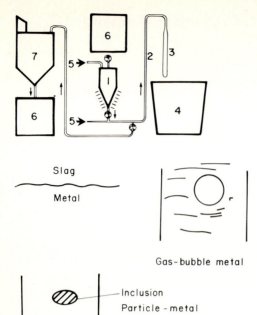

**Fig. 2.1.** Sketch of an injection installation: 1, powder dispenser; 2, transportation hose; 3, lance; 4, ladle or furnace; 5, carrier gas; 6, powder container; and 7, buffer container.

**Fig. 2.2.** Some examples of phase equilibria in injection metallurgy.

1. The "theoretical requirements" represent the minimum reagent needs, and in practice, we will certainly require more, possibly much more than this minimum.
2. The second, perhaps more important point is that thermodynamics can guide us regarding the factors that will influence the phase equilibria, that is, the theoretical reagent requirements. As an example, the reduction in pressure will favor the removal of dissolved gases from melts; the reduction of the oxidizing potential of the slag will favor desulfurization, and the like. It follows that thermodynamic considerations can provide very valuable guidance regarding process selection and process improvements.
3. The third point is that while thermodynamics can guide us regarding the theoretical limits of a given operation, as a practical matter, we will need another set of factors that will define how rapidly these equilibria are approached, that is, process kinetics. Kinetic considerations will be discussed subsequently.

Metallurgical thermodynamics is a complex, extensively documented subject [1, 2, 3, 4, 5, 6], and here we shall confine ourselves to discussing some of the basic principles only.

### 2.2.1 The Principles of Chemical Equilibrium

Let us consider a reaction, such as

$$aA + bB = cC + dD,\tag{2.1}$$

where $A$, $B$, $C$, and $D$ are given chemical species, and $a$, $b$, $c$, and $d$ are the corresponding stoichiometric coefficients. Typical examples of such reactions could include:

$$
\begin{aligned}
\underline{C} + \underline{O} &= CO & a &= 1, b = 1, c = 1, d = 0; \\
2\underline{N} &= N_{2g} & a &= 2, c = 1; \\
2\underline{O} + \underline{Si} &= SiO_2 & a &= 2, b = 1, c = 1; \\
2\underline{P} + 5\underline{O} &= P_2O_5 & a &= 2, b = 5, c = 1.
\end{aligned} \tag{2.2}
$$

(Here the underscore designates a metal phase, and g designates a gas phase, while no subscript denotes the slag phase.) At equilibrium we have

$$
K = \frac{a_C^c a_D^d}{a_A^a a_B^b}, \tag{2.3}
$$

where $K$ is the equilibrium constant, and $a_C$, $a_D$, $a_A$, and $a_B$ are the activities of the respective species. For gases or ideal solutions, the activities would be the same as the mole fractions. However, most solutions of metallurgical interest, that is, alloying elements, such as silicon, carbon, and oxygen, in molten iron or slags are strongly nonideal. For such situations, we have to define the activities as

$$
a_i = f_i x_i, \tag{2.4}
$$

where $x_i$ is the mole fraction of species $i$, and $f_i$ is the activity coefficient, which is strongly composition dependent. This composition dependence of the activity coefficient will be discussed shortly.

The equilibrium constant $K$ is related to $\Delta F^\circ$, the standard free energy charge of the reaction, by

$$
\Delta F^\circ = -RT \ln K, \tag{2.5}
$$

where $R$ is the universal gas constant 1.98 cal/g mole K, and $T$ is the temperature in K. $\Delta F^\circ$ is temperature dependent; information on the temperature dependence of $\Delta F$ is summarized in Table 2.1.

*Example.* Calculate the standard free energy charge associated with the reaction of gaseous oxygen and molten silicon, and hence the equilibrium constant of 1800 K.

*Solution.* for $\underline{Si} + O_{2(g)} = SiO_2$ from Table 2.1 is

$$
\Delta F = -226{,}000 + 47.5 \times 1800
$$

$$
\cong -140{,}000 \text{ cal/mole}
$$

$$
K = \frac{a_{SiO_2}}{a_{Si} a_{O_2}}
$$

$$
-RT \ln K = -140{,}000,
$$

that is,

$$
\ln K = \frac{140{,}000}{1.98 \times 1800}
$$

$$
\cong 39.3, \qquad \text{and further,} \qquad K = 1.17 \times 10^{17},
$$

**Table 2.1.** Standard free energies of some reactions encountered in ferrous metallurgical processes[a]

| Reaction[b] | $\Delta F_r^\circ = A + BT$ (cal.) | | $\pm$ Kcal. | Temperature range (°C) |
|---|---|---|---|---|
| | $-A$ | $B$ | | |
| $2Al\,(s) + \frac{3}{2}O_2\,(g) = Al_2O_3\,(s)$ | 399,500 | 74.71 | 1 | 25–659 |
| $2Al\,(l) + \frac{3}{2}O_2(g) = Al_2O_3\,(s)$ | 402,300 | 77.83 | 1 | 659–1700 |
| $Al\,(s) + \frac{1}{2}N_2\,(g) = AlN\,(s)$ | 75,760 | 25.40 | 1 | 25–659 |
| $Al\,(l) + \frac{1}{2}N_2\,(g) = AlN\,(s)$ | 78,030 | 27.82 | 1 | 659–1700 |
| $4Al\,(s) + 3C\,(s) = Al_4C_3\,(s)$ | 35,080 | −2.60 | 4 | 25–659 |
| $4Al\,(s) + 3C\,(s) = Al_4C_3\,(s)$ | 42,700 | 5.70 | 4 | 659–1700 |
| $2B\,(s) + \frac{3}{2}O_2\,(g) = B_2O_3\,(s)$ | 305,120 | 63.01 | 1 | 25–450 |
| $2B\,(s) + \frac{3}{2}O_2\,(g) = B_2O_3\,(l)$ | 295,630 | 50.41 | 1 | 450–1700 |
| $B\,(s) + \frac{1}{2}N_2\,(g) = BN\,(s)$ | 60,600 | 21.40 | 0.5 | 25–900 |
| $4B\,(s) + C\,(s) = B_4C\,(s)$ | 13,580 | 1.69 | 1 | 25–900 |
| $Ba\,(s) + \frac{1}{2}O_2\,(g) = \text{"BaO"}\,(s)$ | 132,920 | 22.36 | 3.5 | 25–704 |
| $Ba\,(l) + \frac{1}{2}O_2\,(g) = \text{"BaO"}\,(s)$ | 133,400 | 22.87 | 3.5 | 704–1638 |
| $3Ba\,(s) + N_2\,(g) = Ba_3N_2\,(s)$ | 87,000 | 57.4 | 9 | 25–704 |
| $BaO\,(s) + SiO_2\,(s) = BaSiO_3\,(s)$ | 26,800 | 0.1 | 3 | 25–1300 |
| $Be\,(s) + \frac{1}{2}O_2\,(g) = BeO\,(s)$ | 142,900 | 23.13 | 0.5 | 25–1283 |
| $Be\,(l) + \frac{1}{2}O_2\,(g) = BeO\,(s)$ | 142,360 | 23.36 | 1.0 | 1283–1700 |
| $3Be\,(s) + N_2\,(g) = Be_3N_2\,(s)$ | 134,700 | 40.6 | 12 | 25–700 |
| $C\,(s) + 2H_2\,(g) = CH_4\,(g)$ | 21,50 | 26.16 | 1 | 25–2000 |
| $C\,(s) + \frac{1}{2}O_2\,(g) = CO\,(g)$ | 26,760 | −20.98 | 1 | 25–2000 |
| $C\,(s) + O_2\,(g) = CO_2\,(g)$ | 94,60 | −0.27 | 1 | 25–2000 |
| $C\,(s) + \frac{1}{2}S_2\,(g) = CS\,(g)$ | −59,000 | −22.75 | 7 | 1600–1800 |
| $C\,(s) + S_2\,(g) = CS_2\,(g)$ | 3,00 | −1.73 | 1 | 25–1300 |
| $CO\,(g) + \frac{1}{2}S_2\,(g) = COS\,(g)$ | 22,860 | 18.7 | 3 | 25–1200 |
| graphite → diamond | 310 | 1.13 | 0.2 | 25–1200 |
| $Ca\,(s) + \frac{1}{2}O_2\,(g) = CaO\,(s)$ | 150,470 | 24.47 | 1 | 25–850 |
| $Ca\,(l) + \frac{1}{2}O_2\,(g) = CaO\,(s)$ | 152,020 | 25.80 | 1 | 850–1487 |
| $Ca\,(g) + \frac{1}{2}O_2\,(g) = CaO\,(s)$ | 187,980 | 46.21 | 2 | 1487–1700 |
| $Ca\,(s) + \frac{1}{2}S_2\,(g) = CaS\,(s)$ | 129,490 | 22.86 | 1 | 25–850 |
| $Ca\,(l) + \frac{1}{2}S_2\,(g) = CaS\,(s)$ | 131,780 | 24.94 | 1 | 850–1487 |
| $Ca\,(g) + \frac{1}{2}S_2\,(g) = CaS\,(s)$ | 168,360 | 45.72 | 2 | 1487–1700 |
| $3Ca\,(s) + N_2\,(g) = Ca_3N_2\,(s)$ | 105,000 | 50.0 | 10 | 25–850 |
| $Ca\,(s)\,\alpha + 2C\,(s) = CaC_2\,(s)$ | 13,600 | −5.9 | 3 | 25–400 |
| $Ca\,(s)\,\beta + 2C\,(s) = CaC_2\,(s)$ | 11,620 | −8.64 | 3 | 400–850 |
| $Ca\,(l) + 2C\,(s) = CaC_2\,(s)$ | 13,700 | −6.80 | 3 | 850–1487 |
| $Ca\,(g) + 2C\,(s) = CaC_2\,(s)$ | 51,210 | 12.3 | 5 | 1487–1900 |
| $3CaO\,(s) + Al_2O_3\,(s) = Ca_3Al_2O_6\,(s)$ | 3,900 | −6.3 | 2 | 25–1550 |
| $12CaO\,(s) + 7Al_2O_3\,(s) = Ca_{12}Al_{14}O_{33}\,(s)$ | 17,460 | −49.6 | 2 | 25–1500 |
| $CaO\,(s) + Al_2O_3\,(s) = CaAl_2O_4\,(s)$ | 4,570 | −4.1 | 2 | 25–1600 |
| $CaO\,(s) + CO_2\,(g) = CaCO_3\,(s)$ | 40,250 | 34.4 | 1 | 25–880 |
| $2CaO\,(s) + Fe_2O_3\,(s) = Ca_2Fe_2O_5\,(s)$ | 9,200 | −2.33 | 1.5 | 600–1435 |
| $2CaO\,(s) + Fe_2O_3\,(s) = Ca_2Fe_2O_5\,(l)$ | −7,560 | −12.13 | 1.5 | 1435–1600 |
| $4CaO\,(s) + P_2\,(g) + \frac{5}{2}O_2\,(g) = Ca_4P_2O_9\,(s)$ | 563,580 | 144.0 | 3.0 | 1300–1600 |
| $3CaO\,(s) + P_2\,(g) + \frac{5}{2}O_2\,(g) = Ca_3P_2O_8\,(s)$ | 553,350 | 144.0 | 3.0 | 1300–1600 |
| $2CaO\,(s) + SiO_2\,(s) = Ca_2SiO_4\,(s)$ | 30,200 | −1.2 | 2.5 | 25–1400 |
| $CaO\,(s) + SiO_2\,(s) = CaSiO_3\,(s)\,\alpha$ | 21,300 | 0.12 | 1 | 25–1210 |
| $CaO\,(s) + SiO_2\,(s) = CaSiO_3\,(s)\,\beta$ | 19,900 | −0.82 | 2 | 1210–1543 |

**Table 2.1** (*cont.*)

| Reaction[b] | $\Delta F_{\tau}^{\circ} = A + BT$ (cal.) | | $\pm$ Kcal. | Temperature range (°C) |
| --- | --- | --- | --- | --- |
| | $-A$ | $B$ | | |
| $Co$ (s) $+ \frac{1}{2}O_2$ (g) $= CoO$ (s) | 57,380 | 18.65 | 2 | 25–1700 |
| $3Co$ (s) $+ 2O_2$ (g) $= Co_3O_4$ (s) | 206,590 | 82.86 | 3 | 25–1200 |
| $9Co$ (s) $+ 4S_2$ (g) $= Co_9S_8$ (s) | 316,960 | 159,24 | 2 | 25–778 |
| $2Co$ (s) $+ C$ (s) $= Co_2C$ (s) | −3,950 | −2.08 | 5 | 25–900 |
| $2Cr$ (s) $+ \frac{3}{2}O_2$ (g) $= Cr_2O_3$ (s) $\beta$ | 271,300 | 61.82 | 0.5 | 25–1898 |
| $2Cr$ (l) $+ \frac{3}{2}O_2$ (g) $= Cr_2O_3$ (s) $\beta$ | 287,900 | 70.00 | 1 | 1898–2500 |
| $2Cr$ (s) $+ \frac{1}{2}N_2$ (g) $= Cr_2N$ (s) | 24,000 | 11.65 | 5 | 25–1898 |
| $Cr$ (s) $+ \frac{1}{2}N_2$ (g) $= CrN$ (s) | 25,500 | 16.7 | 7.5 | 25–1898 |
| $23Cr$ (s) $+ 6C$ (s) $= Cr_{23}C_6$ (s) | 98,280 | −9.24 | 10 | 25–1400 |
| $7Cr$ (s) $+ 3C$ (s) $= Cr_7C_3$ (s) | 41,800 | −6.1 | 10 | 25–1200 |
| $3Cr$ (s) $+ 2C$ (s) $= Cr_3C_2$ (s) | 20,800 | −4.0 | 10 | 25–1700 |
| $Fe$ (s) $+ \frac{1}{2}O_2$ (g) $=$ "FeO" (s) | 63,200 | 15.47 | 0.5 | 25–1377 |
| $Fe$ (l) $+ \frac{1}{2}O_2$ (g) $=$ "FeO" (l) | 57,070 | 11.60 | 1 | 1537–1700 |
| $3Fe$ (s) $+ 2O_2$ (g) $= Fe_3O_4$ (s) | 265,660 | 76.81 | 1.2 | 25–600 |
| $3Fe$ (s) $+ 2O_2$ (g) $= Fe_3O_4$ (s) | 261,200 | 71.36 | 1.2 | 600–1537 |
| $3Fe$ (l) $+ 2O_2$ (g) $= Fe_3O_4$ (s, l) | 248,240 | 64.42 | 2 | 1537–1700 |
| $2Fe$ (s) $+ \frac{3}{2}O_2$ (g) $= Fe_2O_3$ (s) | 195,450 | 61.38 | 1.5 | 25–680 |
| $2Fe$ (s) $+ \frac{3}{2}O_2$ (g) $= Fe_2O_3$ (s) | 192,800 | 58.30 | 1.5 | 680–1537 |
| $Fe$ (s) $+ \frac{1}{2}S_2$ (g) $= FeS$ (s) $\alpha$ | 37,160 | 15.59 | 1 | 25–140 |
| $Fe$ (s) $+ \frac{1}{2}S_2$ (g) $= FeS$ (s) $\beta$ | 35,910 | 12.56 | 1 | 140–906 |
| $4Fe$ (s) $+ \frac{1}{2}N_2$ (g) $= Fe_4N$ (s) | 1,130 | 9.7 | 1 | 25–600 |
| $3Fe$ (s) $+ \frac{1}{2}P_2$ (g) $= Fe_3P$ (s) | 51,000 | 11.3 | 8 | 25–1170 |
| $3Fe$ (s) $+ C$ (s) $= Fe_3C$ (s) | −6,200 | −5.53 | 1 | 25–190 |
| $3Fe$ (s) $+ C$ (s) $= Fe_3C$ (s) | −6,380 | −5.92 | 1 | 190–840 |
| $3Fe$ (s) $+ C$ (s) $= Fe_3C$ (s) | −2,475 | −2.43 | 1 | 840–1537 |
| "FeO" (s) $+ Al_2O_3$ (s) $= FeAl_2O_4$ (s) | 11,800 | 5.43 | 4 | 25–1377 |
| "FeO" (s) $+ Cr_2O_3$ (s) $= FeCr_2O_4$ (s) | 2,600 | −3.37 | 2 | 25–1377 |
| 2 "FeO" (s) $+ SiO_2$ (s) $= Fe_2SiO_4$ (s) | 7,950 | 3.65 | 2 | 25–1217 |
| 2 "FeO" (s) $+ SiO_2$ (s) $= Fe_2SiO_4$ (l) | −14,880 | −11.49 | 2 | 1217–1377 |
| 2 "FeO" (l) $+ SiO_2$ (s) $= Fe_2SiO_4$ (l) | −3,450 | −4.58 | 2 | 1377–1700 |
| "FeO" (s) $+ TiO_2$ (s) $= FeTiO_3$ (s) | 1,410 | −2.54 | 1 | 900–1377 |
| $H_2$ (g) $+ \frac{1}{2}O_2$ (g) $= H_2O$ (g) | 58,850 | 13.12 | 0.5 | 25–1700 |
| $H_2$ (g) $+ \frac{1}{2}S_2$ (g) $= H_2S$ (g) | 21,580 | 11.80 | 0.5 | 25–1500 |
| $\frac{3}{2}H_2$ (g) $+ \frac{1}{2}N_2$ (g) $= NH_3$ (g) | 12,050 | 26.7 | 2 | 25–700 |
| $Mg$ (s) $+ \frac{1}{2}O_2$ (g) $= MgO$ (s) | 143,000 | 25.91 | 1.5 | 25–650 |
| $Mg$ (l) $+ \frac{1}{2}O_2$ (g) $= MgO$ (s) | 145,830 | 28.10 | 1.5 | 650–1120 |
| $Mg$ (g) $+ \frac{1}{2}O_2$ (g) $= MgO$ (s) | 176,060 | 49.84 | 3 | 1120–1700 |
| $Mg$ (s) $+ \frac{1}{2}S_2$ (g) $= MgS$ (s) | 99,650 | 22.8 | 5 | 25–650 |
| $Mg$ (l) $+ \frac{1}{2}S_2$ (g) $= MgS$ (s) | 101,800 | 25.65 | 5 | 650–1120 |
| $Mg$ (g) $+ \frac{1}{2}S_2$ (g) $= MgS$ (s) | 134,350 | 48.75 | 5 | 1120–1700 |
| $3Mg$ (s) $+ N_2$ (g) $= Mg_3N_2$ (s) | 109,600 | 47.41 | 3 | 25–650 |
| $3Mg$ (l) $+ N_2$ (g) $= Mg_3N_2$ (s) | 115,810 | 54.20 | 3 | 650–1120 |
| $MgO$ (s) $+ CO_2$ (g) $= MgCO_3$ (s) | 28,100 | 40.6 | 3 | 25–700 |
| $3MgO$ (s) $+ P_2$ (g) $+ \frac{5}{2}O_2$ (g) $= Mg_3P_2O_8$ (s) | 501,610 | 144.0 | 3 | 1000–1250 |
| $2MgO$ (s) $+ SiO_2$ (s) $= Mg_2SiO_4$ (s) | 15,120 | 0.0 | 2 | 25–1400 |
| $MgO$ (s) $+ SiO_2$ (s) $= MgSiO_3$ (s) | 8,900 | 1.1 | 1 | 25–1300 |

**Table 2.1** (*cont.*)

| Reaction[b] | $\Delta F_r^\circ = A + BT$ (cal.) | | $\pm$ Kcal. | Temperature range (°C) |
| | $-A$ | $B$ | | |
|---|---|---|---|---|
| Mn (s) + $\frac{1}{2}O_2$ (g) = MnO (s) | 92,490 | 17.87 | 1.5 | 25–1244 |
| Mn (l) + $\frac{1}{2}O_2$ (g) = MnO (s) | 97,360 | 21.12 | 1.5 | 1244–1700 |
| 3Mn (s) + 2$O_2$ (g) = $Mn_3O_4$ (s) | 330,750 | 83.61 | 1.5 | 25–1244 |
| 3Mn (l) + 2$O_2$ (g) = $Mn_3O_4$ (s) | 339,000 | 89.03 | 1.5 | 1244–1700 |
| Mn (s) + $\frac{1}{2}S_2$ (g) = MnS (s) | 65,000 | 16.21 | 1.5 | 25–1244 |
| Mn (l) + $\frac{1}{2}S_2$ (g) = MnS (s) | 69,010 | 18.86 | 2 | 1244–1530 |
| Mn (l) + $\frac{1}{2}S_2$ (g) = MnS (l) | 62,770 | 15.40 | 2 | 1530–1700 |
| 3Mn (s) + C (s) = $Mn_3C$ (s) | 3,330 | −0.26 | 3 | 25–740 |
| MnO (s) + $SiO_2$ (s) = $MnSiO_3$ (s) | 5,920 | 3.0 | 4 | 25–1300 |
| 2Mo (s) + $\frac{1}{2}N_2$ (g) = $Mo_2N$ (s) | 15,250 | 13.25 | 3 | 25–1000 |
| 2Mo (s) + C (s) = $Mo_2C$ (s) | 6,700 | 0.0 | 8 | 25–1000 |
| Ni (s) + $\frac{1}{2}O_2$ (g) = NiO (s) | 58,450 | 23.55 | 2 | 25–1452 |
| Ni (l) + $\frac{1}{2}O_2$ (g) = NiO (s) | 62,650 | 25.98 | 3 | 1452–1900 |
| Ni (s) + $\frac{1}{2}S_2$ (g) = NiS (s) | 39,980 | 17.20 | 3 | 400–580 |
| 3Ni (s) + C (s) = $Ni_3C$ (s) | −8,100 | −1.70 | 3 | 25–700 |
| 2P (s, l) + $\frac{5}{2}O_2$ (g) = $P_2O_5$ (s) | 361,300 | 113.6 | 10 | 25–280 |
| $\frac{1}{2}S_2$ (g) + $O_2$ (g) = $SO_2$ (g) | 86,520 | 17.48 | 1 | 25–1700 |
| Si (s) + $\frac{1}{2}O_2$ (g) = SiO (g) | 22,600 | −19.71 | 3 | 25–1413 |
| Si (l) + $\frac{1}{2}O_2$ (g) = SiO (g) | 36,150 | −11.51 | 3 | 1413–1700 |
| Si (s) + $O_2$ (g) = $SiO_2$ (s) α cristobalite | 215,600 | 42.26 | 3 | 25–250 |
| Si (s) + $O_2$ (g) = $SiO_2$ (s) β cristobalite | 214,400 | 40.32 | 3 | 250–1413 |
| Si (l) + $O_2$ (g) = $SiO_2$ (s) β cristobalite | 226,500 | 47.50 | 3 | 1413–1700 |
| 3Si (s) + 2$N_2$ (g) = $Si_3N_4$ (s) | 172,700 | 75.2 | 5 | 25–1413 |
| 3Si (l) + 2$N_2$ (g) = $Si_3N_4$ (s) | 209,000 | 96.8 | 2 | 1413–1700 |
| Si (s) + C (s) = SiC (s) β | 13,000 | 0.73 | 2 | 25–1413 |
| Si (l) + C (s) = SiC (s) β | 25,100 | 7.91 | 2 | 1413–1700 |
| Ti (s) + $O_2$ (g) = $TiO_2$ (s) | 224,500 | 42.33 | 2 | 25–1700 |
| Ti (s) + $\frac{1}{2}N_2$ (g) = "TiN" (s) | 80,550 | 22.45 | 2 | 25–1300 |
| Ti (s) + C (s) = "TiC" (s) | 45,000 | 2.79 | 3 | 25–1700 |
| 2V (s) + $\frac{3}{2}O_2$ (g) = $V_2O_3$ (s) | 291,350 | 56.49 | 6.5 | 25–1700 |
| 2V (s) + $\frac{5}{2}O_2$ (g) = $V_5O_3$ (s) | 371,400 | 100.92 | 5.5 | 25–670 |
| 2V (s) + $\frac{5}{2}O_2$ (g) = $V_2O$ (l) | 349,250 | 77.46 | 5.5 | 670–1700 |
| V (s) + $\frac{1}{2}N_2$ (g) = "VN"[b] (s) | 41,650 | 19.35 | 10 | 25–1300 |
| V (s) + C (s) = "VC" (s) | 12,500 | 1.6 | 10 | 25–1700 |
| W (s) + C (s) = WC (s) | 9,100 | 0.4 | 3 | 25–1700 |
| Zr (s) + $O_2$ (g) = $ZrO_2$ (s) α | 260.730 | 44.7 | 4 | 25–1205 |
| Zr (s) + $O_2$ (g) = $ZrO_2$ (s) β | 258,170 | 42.87 | 4 | 1205–1700 |
| Zr (s) + $\frac{1}{2}N_2$ (g) = "ZrN" (s) | 87,470 | 22.72 | 3 | 25–1300 |
| Zr (s) + C (s) = "ZrC" (s) | 44,100 | 2.2 | 3 | 25–1900 |

[a] Most of the data are taken from the following sources: (1) J.P. Coughlin, U.S. Bureau of Mines Bull. No. 542 (1954); (2) O. Kubaschewski and E. Ll. Evans, "Metallurgical Thermochemistry," Pergamon, London (1958); and (3) J.F. Elliott and M. Gleiser, "Thermochemistry for Steelmaking," Vol. I, Addison-Wesley, MA (1960).
[b] Notations: (g), gas; (l), liquid; (s), solid; " ", nonstoichiometric compounds.

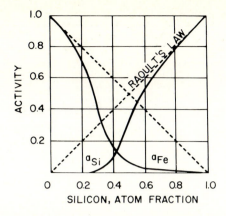

**Fig. 2.3.** The activities of iron–silicon melts at 1600°C, after Chipman *et al.*, cited by Fruehan [6].

which means that equilibrium is very much in favor of the forward reaction, that is, $SiO_2$ formation. In fact, this could have been deduced already from the value of $\Delta F°$, as a strongly negative value indicates automatically that $K = \exp(-\Delta F°/RT)$ becomes strongly positive, again, indicating that the equilibrium lies on the side of the reaction products.

Another particular case of chemical equilibrium is the partitioning of elements between a gaseous and a molten phase. Thus, we have

$$P_i = P_{i,o} f_i x_i, \tag{2.6}$$

where $P_i$ is the partial pressure of a given species in the gas phase; $f_i$ is the activity coefficient; and $P_{i,o}$ is the vapor pressure of the pure substance. Often Eq. (2.6) may be approximated by

$$P_i = K x_i, \tag{2.7}$$

which is called Henry's Law, where $K$ is Henry's constant.

It should be stressed to the reader that in the majority of applications, the solutions considered are nonideal, and for this reason we must have information on the activity coefficients before we can proceed further. This is illustrated in Fig. 2.3, showing how the activity of silicon in iron–silicon melts would vary with composition. The broken line, depicting Raoult's Law behavior, corresponds to ideal solutions, that is, to a unit activity coefficient, which would give a highly erroneous result. Information on activity coefficients and on the way they may be estimated is available in Ref. 1–6. Let us turn our attention to some practical applications of chemical equilibrium considerations.

### 2.2.1.1 Deoxidation of Equilibria

Deoxidation reactions may be generally represented in this fashion:

$$m\underline{M} + o\underline{O} = M_m O_o \tag{2.8}$$

where $M$ is the deoxidant, such as silicon or aluminum.

**Table 2.2.** Deoxidation solubility products in liquid iron

| Equilibrium constant $K^a$ | Composition range | $K$ (at 1600°C) | $\log K$ |
|---|---|---|---|
| $[a_{Al}]^2[a_O]^4$ | <1 ppm Al | $1.1 \times 10^{-15}$ | $-\dfrac{71{,}600}{T} + 23.28$ |
| $[a_{Al}]^2[a_O]^3$ | >1 ppm Al | $4.3 \times 10^{-14}$ | $-\dfrac{62{,}780}{T} + 20.41$ |
| $[a_B]^2[a_O]^3$ | | $1.3 \times 10^{-8}$ | |
| $[a_C][a_O]/P_{CO}$ | >0.02% C | $2.0 \times 10^{-3}$ | $-\dfrac{1168}{T} - 2.07$ |
| $[a_{Cr}]^2[a_O]^4$ | <3% Cr | $4.0 \times 10^{-6}$ | $-\dfrac{50{,}700}{T} + 21.70$ |
| $[a_{Cr}]^2[a_O]^3$ | >3% Cr | $1.1 \times 10^{-4}$ | $-\dfrac{40{,}740}{T} + 17.78$ |
| $[a_{Mn}][a_O]$ | >1% Mn | $5.1 \times 10^{-2}$ | $-\dfrac{14{,}450}{T} + 6.43$ |
| $[a_{Si}][a_O]^2$ | >20 ppm Si | $2.2 \times 10^{-5}$ | $-\dfrac{30{,}410}{T} + 11.59$ |
| $[a_{Ti}]^3[a_O]^5$ | 0.01–0.25% Ti | $7.9 \times 10^{-17}$ | |
| $[a_v]^2[a_O]^4$ | <0.1% V | $8.3 \times 10^{-8}$ | $-\dfrac{48{,}060}{T} + 18.61$ |
| $[a_v]^2[a_O]^3$ | >0.3% V | $3.5 \times 10^{-6}$ | $-\dfrac{43{,}200}{T} + 17.52$ |

[a] Activities are chosen such that $a_M \equiv \%M$ and $a_O \equiv \%O$ when $\%M \to O$.

Then, we may write

$$K_M = \frac{a_{MO}}{a_M^m a_O^o}. \tag{2.9}$$

A listing of equilibrium constants for deoxidation is given in Table 2.2.

2.2.1.1.1 Complex Deoxidation

When more than one deoxidant is used, such as Si–Mn or Si–Ca, two possible benefits may be derived.

1. The soluble oxygen may be reduced, because the activity of the complex oxide formed is reduced.
2. The deoxidation products formed may be in the molten state, and thus may be more readily floated out.

As an example, Fig. 2.4 shows the equilibrium oxygen activity in iron–aluminum alloys as a function of the CaO content of the slag. It is seen that above 37 wt. %, the presence of CaO will markedly reduce the oxygen activity at equilibrium.

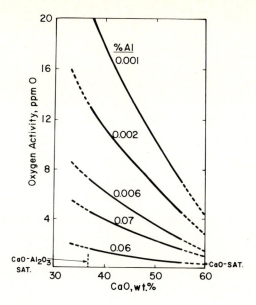

**Fig. 2.4.** The activity of oxygen in Fe–Al alloys in equilibrium with a CaO–Al O slag at 1600°C, cited by Fruehan [6].

### 2.2.1.2 Desulfurization Equilibria

CaO, Ca, or other calcium compounds such as calcium carbide are the principal desulfurizing agents, and the corresponding reaction may be put in the following form:

$$CaO + \underline{S} = CaS + \underline{O};$$ (2.10)

thus, the equilibrium constant may be expressed as

$$K = \frac{a_{CaS} a_O}{a_{CaO} a_S},$$ (2.11)

which immediately shows that the lower the oxygen activity, the lower will be the equilibrium sulfur content of the steel. This clearly underlines the need to deoxidize the steel first, before desulfurization.

Figure 2.5 shows the partitioning of sulfur between slag and metal in CaO–Al$_2$O$_3$ slags at two temperatures, as a function of the aluminum content of the slag.

It is seen that the partition ratio is increased with increasing aluminum (because this corresponds to a lowering of the oxygen activity); furthermore, lower temperatures will tend to favor desulfurization equilibria. However, as we shall discuss subsequently, kinetics are favored at higher temperatures, so that here some accommodation will have to be reached.

An alternative way of expressing desulfurization equilibria is through the use of the sulfide capacity concept [7].

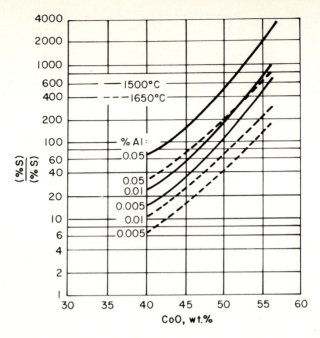

**Fig. 2.5.** Sulfur partitioning between slag and metal for Fe–Al alloys in equilibrium with $CaO–Al_2O_3$ slags, cited by Fruehan [6].

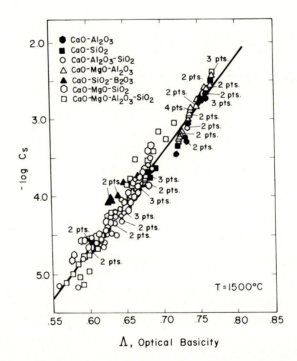

**Fig. 2.6.** The relationship between the optical basicity and the sulfide capacity, after Sommerville [7].

**Table 2.3.** Relative basicities of oxides

| By carbonate decomposition, 1600°C [14] | By bond strength [15] | On optical basicity scale | $\Lambda$ |
|---|---|---|---|
| BaO | $K_2O$ | $K_2O$ | 1.40 |
| $Li_2O$ | $Na_2O$ | $Na_2O$ | 1.15 |
| CaO | $Li_2O$ | BaO | 1.15 |
| PbO | BaO | SrO | 1.07 |
| CdO | PbO | $Li_2O$ | 1.00 |
| MgO | SrO | CaO | 1.00 |
| AgO | CaO | MgO | 0.78 |
| ZnO | MgO | $ZrO_2$ | 0.69 |
| $TiO_2$ | $ZrO_2$ | $TiO_2$ | 0.61 |
| $SiO_2$ | BeO | $Al_2O_3$ | 0.605 |
| $B_2O_3$ | $Al_2O_3$ | BeO | 0.57 |
| $P_2O_5$ | $TiO_2$ | $SiO_2$ | 0.48 |
| | $B_2O_3$ | $B_2O_3$ | 0.42 |
| | $SiO_2$ | $P_2O_5$ | 0.40 |
| | $P_2O_5$ | | |

The sulfide capacity of the slag may be defined as

$$C_S = \text{wt. \%S} \left( \frac{P_{O_2}}{P_{S_2}} \right)^{1/2} \tag{2.12}$$

or

$$C_s' = \text{wt. \%S} \left( \frac{a_O}{a_S} \right). \tag{2.13}$$

Somerville has shown that the sulfide capacity of slags may be readily related to the optical basicity of the slag. Figure 2.6 shows a plot of the sulfide capacity as a function of the optical basicity, $\Lambda$.

Here, $\Lambda$ is defined as

$$\Lambda = \sum \Lambda_i X_i, \tag{2.14}$$

where the $\Lambda_i$ values corresponding to the various oxides are listed in Table 2.3. Once $\Lambda_i$ is known, the sulfur partitioning may be calculated with the aid of Eq. (2.15):

$$\log \frac{\%S}{[a_s]} = \frac{21920 - 54640\Lambda}{T} + 43.6\Lambda - 23.9 - \log[a_o]. \tag{2.15}$$

### 2.2.1.3 Sulfide Shape Control

When steel is strongly deoxidized at normally encountered sulfur and manganese levels, MnS inclusions may form. These inclusions have a relatively low melting point and on rolling may lead to the formation of string-type inclusions, which have a serious adverse effect on the transverse mechanical properties.

In order to avoid the occurrence of such behavior, the inclusions may be modified by the addition of calcium or rare earths, such as cerium and the like, which will lead to the formation of complex oxi-sulfides, thus avoiding the previously described problems.

### 2.2.1.4 Alumina Shape Control

When steel is deoxidized with aluminum, alumina inclusions, which have a high melting point, will form. These inclusions may pose serious problems in blocking nozzles in continuous casting, and also in the mechanical properties. Alumina inclusions may be modified through the addition of calcium or rare earths, which again will result in the formation of complex oxides.

The thermodynamic equilibrium considerations pertaining to these systems are discussed in greater detail in Ref. 4–6, and the practical aspects of modification of inclusions are discussed in Chapter 3 of this book.

### 2.2.1.5 Dephosphorization

The chemical reaction involving dephosphorization may be written as

$$2\underline{P} + 5\underline{O} = P_2O_5$$

or

$$2\underline{P} + 5\underline{O} + nCaO = P_2O_5 nCaO$$

or

$$\tfrac{5}{4}\underline{P} + Na_2CO_3 = Na_2O + \tfrac{2}{5}P_2O_5 + [C].$$

Indeed, dephosphorization is usually carried out by the injection of lime, $CaF_2$, or $Na_2CO_3$; see a brief discussion of these issues in Chapter 3, which deals with injection practice.

Dephosphorization poses an interesting problem, because in contrast to desulfurization, high oxygen content of the metal and high temperatures will favor phosphorus removal.

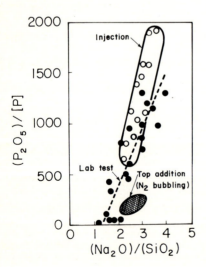

**Fig. 2.7.** Dephosphorization of hot metal by injection or addition to the top of the metal phase.

In a recent paper, Emi and Iida [8] have presented a very useful discussion of new processes that may enable the simultaneous desulfurization and dephosphorization of melts by sodium carbonate injection. Figure 2.7 shows some laboratory scale dephosphorization results that look very promising.

### 2.2.1.6 Degassing

Equilibrium considerations can also be very helpful in assessing the theoretical limits of hydrogen, nitrogen, and oxygen removal due to purging by inert gases such as argon by using a simple relationship of the following type:

$$\frac{dx_i}{dt} N = V(x_i - x_{i,\text{eq}}), \tag{2.16}$$

where $N$ is the total moles of steel to be degassed, $x_i$ is the mole fraction of the constituent to be removed, $V$ is the volume of the purge gas required; and $x_{i,\text{eq}}$. is the equilibrium mole fraction of species $i$. Such degassing is of importance in bottom blown steelmaking processes, and also in ladle metallurgy operations involving gas injection.

Many calculations of this type have been reported in the literature, notably by Turkdogan [9–11] and other; some typical examples are shown in Figs. 2.8 and 2.9.

Two remarks should be made here. One of these is that the form of Eq. (2.16), that is, the logarithmic expression, clearly indicates that a close approach to equilibrium will require large purge gas volumes. The second is that while the equilibrium assumption is quite good for deoxidation, via the $C + O = CO$ reaction or hydrogen removal through $2H = H_2$, in the case of nitrogen removal, there are serious kinetic limitations that introduce an additional barrier so that nitrogen removal by argon purging is hardly ever a practical possibility.

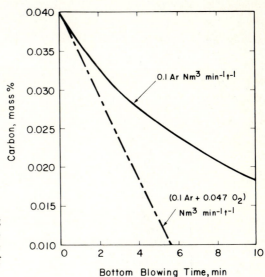

**Fig. 2.8.** Decarburization of steel during bottom blowing with argon or argon oxygen calculated for the indicated gas flow rates and an initial oxygen content of 700 ppm.

We should remark here that through the ready availability of computerized thermodynamic data bases, even quite complex thermodynamic equilibrium calculations may now be carried out quite easily and rapidly.

An excellent review of the available thermodynamic packages has been recently presented by Bjorkmann and Jacobson [12].

## 2.3 Gas and Solids Delivery

### 2.3.1 Gas Delivery Systems

Gases, primarily argon, are injected into ladles in order to promote mixing, effect degassing, and introduce solid particles. As sketched in Fig. 2.10, gases may be intro-

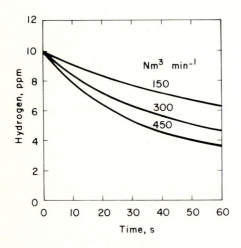

**Fig. 2.9.** Rate of dehydrogenization of steel with inert gas purging in a Q-BOP calculated for a 200-ton heat for various gas flow rates.

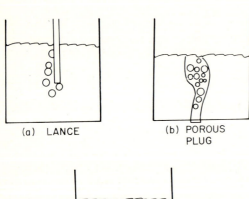

**Fig. 2.10.** Gas injection methods.

duced into melts either through lances, tuyeres, or porous plugs. In this section we shall confine our interest to some basic aspects of gas injection only.

When a gas is injected into a liquid through an orifice, several flow regimes may exist. At low gas flow rates, discrete bubbles are formed [see Fig. 2.12(b)], the size of which is determined by a balance between inertial and surface tension forces. The bubble departure diameters ($d_B$) may be estimated with the aid of an expression suggested by Sano and Mori [13]:

$$d_B = \left[ \frac{3\sigma d_{n,0}}{\rho g} N'_{Sc} + \left( \frac{9\sigma^2 d_{n,0}}{\rho^2 g^2} + \frac{10 Q^2 d_{n,0}}{g} \right)^{1/2} \right]^{1/3}, \tag{2.17}$$

where $d_B$ is the bubble diameter, $\sigma$ is the surface tension, $d_{n,0}$ is the outer diameter of the nozzle, $N'_{Sc}$ is the Schmidt number, $\rho$ is the gas density, and $Q$ is the volumetric gas flow rate. Figure 2.11 shows a comparison between experimental measurements and predictions based on Eq. (2.17).

For higher gas velocities, which will be the case in all ladle metal-lurgy applications, we shall be in the jetting regime, so that a plume, rather than discrete bubbles, is being formed, as shown in Fig. 2.12(a).

In case of horizontal gas injection through tuyeres, for example, in the AOD process sketched in Fig. 2.13, one of the important considerations is to ensure adequate penetration of the gas stream; if this is not achieved, excessive wall erosion may occur, and at the same time the gas is not brought into good contact with the melt.

The penetration of horizontal gas streams into melts has been modeled [14], and

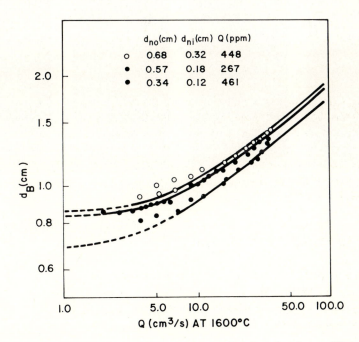

**Fig. 2.11.** Comparison of the experimentally measured bubble sizes in molten iron at different impurity levels with predictions based on Eq. (2.17).

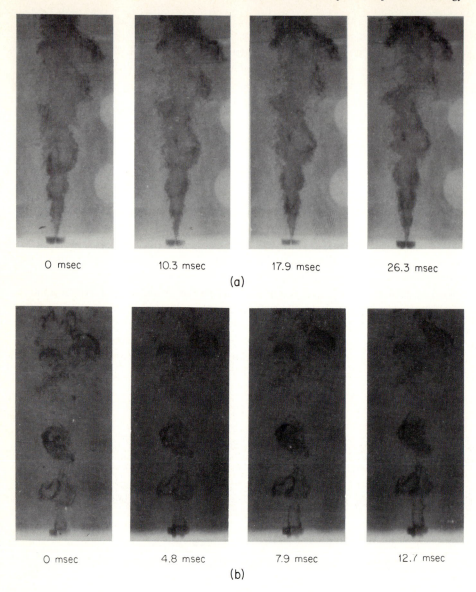

Fig. 2.12. Photographs depicting a gas stream issuing from a nozzle into water: (a) jet formation at high gas flow rates; and (b) the formation of discrete gas bubbles.

**Fig. 2.13.** A schematic diagram of the AOD process.

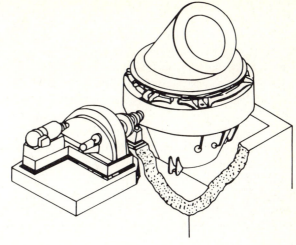

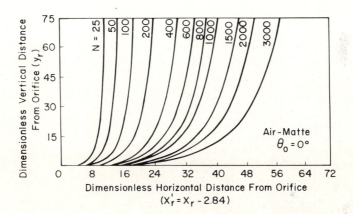

**Fig. 2.14.** The effect of the modified Froude number on the trajectory of an air jet injected into the matte phase in a copper converter. It is seen that the higher the value of $N$, the deeper is the jet penetration.

some computed results are shown in Fig. 2.14, where it is seen that the Froude number, defined as

$$N_{Fr} = \frac{\rho_G U_0^2}{g(\rho_L - \rho_G)d_0}, \tag{2.18}$$

is a key parameter here, where $\rho_G$ is the gas density, $U_0$ is the linear gas velocity at the nozzle exit, $\rho_L$ is the density of the melt, and $d_0$ is the nozzle diameter. The higher the gas velocity, the deeper will be the penetration.

As a matter of fact, in the AOD process, sonic velocity is maintained at the nozzle in order to ensure adequate penetration; furthermore, there seems to be a consensus in the literature that linear gas velocities in the range of 0.6–0.8 Mach are needed in horizontal nozzles in order to ensure that no "backwash" of the metal will occur, which could lead to nozzle blockage [15, 16].

### 2.3.1.1 Particle Delivery

As discussed in Chapter 3, which deals with the practical aspects of injection metallurgy, the delivery of particles to the injection nozzle involves their fluidization and subsequent transportation by means of a carrying gas. The nature of fluidization is perhaps best illustrated by considering the upward flow of a fluid through a packed bed of solids, as sketched in Fig. 2.15, progressively increasing fluid flow rates [17].

It is seen that at low fluid velocities the system behaves like a packed bed; thus, the relationship between the pressure drop across the bed and the flow rate is given by the Ergun equation. As the fluid velocity is increased, this will cause a corresponding increase in the pressure drop across the bed until a state is reached when the pressure drop across the bed equals the weight of the bed. At this point, as illustrated in Fig. 2. 15(b), the particles in the bed will become rearranged so as to offer less resistance to the flow and the bed expands to attain the loosest possible packing. At higher fluid velocities the bed expands further, the particles become freely suspended in the gas stream, and the bed attains the fluidized state. From a macroscopic point of view, gas–solid fluidized beds may be regarded as well-stirred, boiling liquids; the liquidlike properties of fluidized beds are shown in Fig. 2.16.

The onset of fluidization sketched in Fig. 2.15 may be represented conveniently through the use of the fluidization curve shown in Fig. 2.17. In this graph, the pressure drop across the bed is plotted against gas velocity as a logarithmic scale;

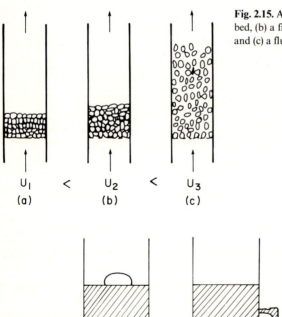

**Fig. 2.15.** A schematic representation of (a) a fixed bed, (b) a fixed bed having its maximum voidage, and (c) a fluidized bed.

$U_1 \quad < \quad U_2 \quad < \quad U_3$
(a)       (b)       (c)

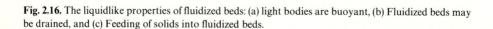

(a)        (b)        (c)

**Fig. 2.16.** The liquidlike properties of fluidized beds: (a) light bodies are buoyant, (b) Fluidized beds may be drained, and (c) Feeding of solids into fluidized beds.

here line AB corresponds to the pressure drop across the fixed bed before fluidization takes place, and region BC represents the rearrangement of the bed to provide the minimum fluidization voidage, when the pressure drop across the bed equals the weight of the bed. The bed is considered fluidized at point C and the corresponding voidage $\varepsilon_{mf}$ is the minimum voidage of the fluidized bed. The broken line DE shows the pressure drop–gas velocity relationship for a packed bed, with an initial voidage of $\varepsilon_{mf}$. The ABC curve represents the ideal behavior of gas–solid fluidized beds. There are many systems that do not follow this pattern exactly, channelling and slugging being typical examples. In a physical sense, channelling corresponds to the preferential flow of gas through certain vertical sections of the bed; thus, under these conditions, a part of the bed may become fluidized while the remainder stays in a packed state.

At the onset of fluidization, the pressure drop across the bed equals the weight of the bed; this relationship may be used to calculate the minimum fluidization velocity:

$$\Delta P = Z_{mf}(1 - \varepsilon_{mf})g(\rho_p - \rho), \tag{2.19}$$

which states that the pressure drop is equal to the weight of the bed.

When a stream of bubbles rises through a bed, then the velocity of an individual bubble is affected by the presence of the other bubbles. For these circumstances, Davidson and Harrison [18] suggested the following expression for estimating the rising velocity of the bubbles:

$$U_n = U_0 - U_{mf} + 0.711(gd_B)^{1/2}. \tag{2.20}$$

Gas bubbles have a major influence on the operation of fluidized systems.

The presence of bubbles is responsible for the good solids mixing, which is a desirable feature. On the other hand, the gas passing through in the form of bubbles is not brought into intimate contact with the bed, as is the gas contained in the continuous, that is, emulsion, phase of the system [19, 20].

### 2.3.2 Pneumatically Conveyed Systems

In order to illustrate the nature of pneumatic conveying, let us reexamine Figs. 2.15 and 2.17, which were given previously. It was shown that when a gas is blown verti-

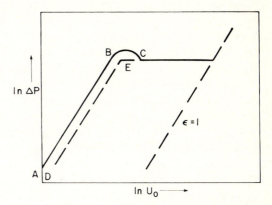

**Fig. 2.17.** The fluidization curve.

cally through a packed bed of solids, at a given gas velocity the bed expands so as to attain the loosest possible packing, and then on a subsequent increase in the gas velocity the bed becomes fluidized. Further increases in the volumetric flow rate of the gas cause part of this gas to pass through the system in the form of bubbles. We also noted that, if the superficial gas velocity exceeds the terminal falling velocity of some of the particles, these fines are elutriated at some finite rate.

However, if a gas were blown through the system at a rate that exceeds the terminal falling velocity of the bulk of the particles contained in the bed, the particles would be swept out of the bed; alternatively, if operated at these rates, particles could be conveyed through conduits.

Such systems are usually called "pneumatically conveyed" and may operate either through horizontal or vertical conduits. The volumetric ratio (gas flow rate)/(solids flow rate) usually ranges between 20–80 in these systems.

The actual purpose of pneumatically conveyed systems is, as the name implies, to transport material; in some cases, however, chemical reactions are also being carried out. These latter systems are termed "transfer line reactors."

The two principal questions that need answering in the design and operation of pneumatic transport installations is the ratio of the gas to solids flow rate required to bring about transport and the pressure drop needed to provide the required gas flow rate.

### 2.3.3 The Saltation Velocity and Choking Phenomena

It has been found experimentally that in the horizontal pneumatic transport of solids there must be a minimum linear gas velocity for a given set of operating conditions (conduit, size, solids flow rate, etc.) to prevent settling or saltation of the solid particles. Such settling or saltation would, of course, result in the blockage of the conduit.

The phenomenon of saltation is illustrated in Fig. 2.18, on a plot of the pressure gradient in the horizontal conduit against the linear gas velocity for various values of the solids flow rate $G_s$. The line corresponding to $G_s = 0$ corresponds to the frictional loss due to the gas flow alone, while the curves designated $G_{s1}$, $G_{s2}$, etc., show the behavior of the system for progressively increasing solids flow rates. Upon considering the curve corresponding to $G_{s_1}$, starting at point C, it is seen that as the linear gas velocity is reduced (at a constant solids loading) pressure drop will also be caused to drop, until point D is reached. At this stage, the solid particles will begin to settle, and a dynamic equilibrium will be established between this settled layer and the mixture above. At point D, the resistance to flow increased very sharply, and a further reduction in the linear gas velocity would cause a further increase in the flow rates in Fig. 2.18; qualitatively similar behavior is observed for higher solids loading except for the fact that the "break" occurs at higher gas velocities. The critical gas velocity, corresponding to point D, is termed the saltation velocity $U_{c,s}$. As a practical matter, a horizontally conveyed system has to be operated above this saltation velocity.

The information available for calculating the saltation velocity is far from adequate; nonetheless, a first-order estimate may be made by following a procedure suggested by Zenz and Othmer [21]. Figure 2.19(a) shows a plot of

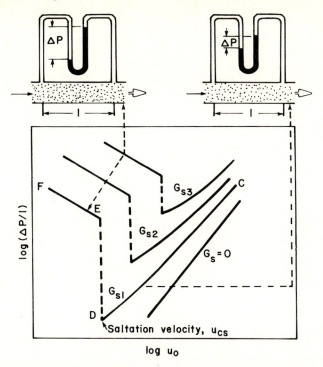

**Fig. 2.18.** The behavior of solids conveyed by a gas stream in a horizontal pipe. The ordinate gives the pressure gradient, while the abscissa designates the linear gas velocity.

$$\frac{U_{c,s,m}}{4g\mu(\rho_p - \rho)^{1/3}/9\rho^2 R_c^{0.4}}$$

against

$$\frac{d_{p,m}}{(3\mu^2/rg\rho(\rho_p - \rho))^{1/3}},$$

from which the saltation velocity of uniformly sized solids of diameter $d_{p,m}$ may be calculated. Here, $R_c$ is the radius of the conduit. For a mixture of sizes, we compute $U_{c,s,m1}$ and $U_{c,s,m2}$, the velocity required to convey the smallest and the largest particles in the mixture, and then find $n$, the slope of the straight line connecting these two points in Fig. 2.19(a). The value of $n$ thus obtained may then be used in conjunction with Fig. 2.19(b) to calculate the critical saltation velocity. An alternative and perhaps more satisfactory way of sizing pipelines for the horizontal conveying of solids is through the use of the minimum safe velocities, listed in Table 2.4 for a variety of systems.

In vertical pneumatic conveying, the concept of choking is somewhat analogous to the saltation phenomenon. The choking phenomenon for the vertical conveying of a lean gas–solid mixture is sketched in Fig. 2.20 on a plot of the linear gas velocity against the (vertical) pressure gradient. Here, again (as in the case of Fig. 2.18), $G_s = 0$ designates the flow of the solids-free gas stream, while $G_{s1}$ and $G_{s2}$ denote progressively higher solids loading.

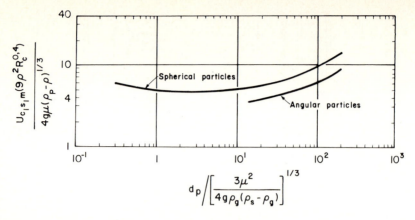

**Fig. 2.19(a).** Correlation for the saltation velocity of uniformly sized solids, after Zenz and Othmer [21].

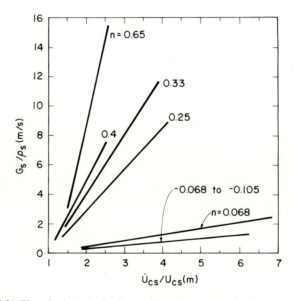

**Fig. 2.19(b).** The saltation velocity for a solids mixture, after Zenz and Othmer [21].

It is seen, on examining curve $G_{s1}$ for example, starting from point C, that the pressure gradient (i.e., the overall pressure drop) decreases as the linear gas velocity is lowered at constant solids loading, down to point D. From this point onward, any further decrease in the linear gas velocity will cause a rapid decrease in the voidage, with a corresponding increase in the static head and the pressure drop, until point E is reached, when the bulk density of the bed becomes too great and the bed collapses into a slugging mode. The superficial velocity at point E is called the choking velocity $U_{ch}$.

It follows from the foregoing that any vertical conveying system has to be operated above the choking velocity. The conditions when a lean gas–solid mixture collapses into a choking mode may be represented by the following expression:

**Table 2.4.** Safe values for pneumatic conveying[a]

| Material | Average bulk density (kg/m³) | Approximate size (1000 μm = 1 mm) | Minimum safe air velocity (m/s) | | Maximum safe density for flow (g/cm³) | |
|---|---|---|---|---|---|---|
| | | | Horizontal | Vertical | Horizontal | Vertical |
| Coal | $0.72 \times 10^3$ | < 12.7 mm | 15.3 | 12.2 | 0.012 | $0.016 \times 10^3$ |
| Coal | $0.72 \times 10^3$ | < 6.35 | 12.2 | 9.2 | 0.016 | $0.024 \times 10^3$ |
| Wheat | $0.75 \times 10^3$ | 4.76mm | 12.2 | 9.2 | 0.024 | $0.032 \times 10^3$ |
| Cement | $1.04$–$1.44 \times 10^3$ | 95% < 88 μm | 7.6 | 1.5 | 0.16 | $0.96 \times 10^3$ |
| Pulverized coal | $0.56 \times 10^3$ | 100% < 380μm, 75% < 76 μm | 4.6 | 1.5 | 0.11 | $0.32 \times 10^3$ |
| Pulverized ash | $0.72 \times 10^3$ | 90% < 150 μm | 4.6 | 1.5 | 0.16 | $0.48 \times 10^3$ |
| Bentonite | $0.77$–$1.04 \times 10^3$ | 95% < 76 μm | 7.6 | 1.5 | 0.16 | $0.48 \times 10^3$ |
| Silica flour | $0.80$–$0.96 \times 10^3$ | 95% < 105 μm | 6.1 | 1.5 | 0.08 | $0.32 \times 10^3$ |
| Phosphate rock | $1.28 \times 10^3$ | 90% < 152 μm | 9.2 | 3.1 | 0.11 | $0.32 \times 10^3$ |
| Common salt | $1.36 \times 10^3$ | 5% < 152 μm | 9.2 | 3.1 | 0.08 | $0.24 \times 10^3$ |
| Soda ash (light) | $0.56 \times 10^3$ | 66% < 105 μm | 9.2 | 3.1 | 0.08 | $0.24 \times 10^3$ |
| Soda ash (dense) | $1.04 \times 10^3$ | 50% < 177 μm | 12.2 | 3.1 | 0.048 | $0.16 \times 10^3$ |
| Sodium sulphate | $1.28$–$1.44 \times 10^3$ | 100% < 500μm, 55% < 105 μm | 12.2 | 3.1 | 0.08 | $0.24 \times 10^3$ |
| Ground bauxite | $1.44 \times 10^3$ | 100% < 105 μm | 7.6 | 1.5 | 0.13 | $0.64 \times 10^3$ |
| Alumina | $0.93 \times 10^3$ | 100% < 105 μm | 7.6 | 1.5 | 0.096 | $0.48 \times 10^3$ |
| Magnesite | $1.60 \times 10^3$ | 90% < 76 μm | 9.2 | 3.1 | 0.16 | $0.48 \times 10^3$ |
| Uranium dioxide | $3.52 \times 10^3$ | 100% < 152 μm, 50% < 76 μm | 18.3 | 6.1 | 0.16 | $0.96 \times 10^3$ |

[a] Adapted from Zentz and Othmer [21].

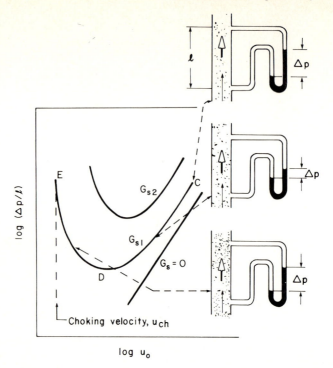

**Fig. 2.20.** The behavior of a lean gas–solid mixture in vertical conveying.

$$G_{s,ch} = \rho_p(1 - \varepsilon_{ch})(U_{ch} - U_t), \tag{2.21}$$

where $G_{s,ch}$ is the solids mass velocity corresponding to choking, $\varepsilon_{ch}$ is the void fraction corresponding to choking, and $U_t$ is the terminal falling velocity of the solid particles. Zenz and Othmer suggested that $\varepsilon_{ch}$ does not depend very much on the size of the solid particles, but that it is a strong function of the solids density. These authors also found that for closely sized particles, the choking velocity can be approximated by the saltation velocity $U_{c,s}$.

Table 2.4 also contains useful information on the safe values of the linear gas velocity for the vertical pneumatic conveying of solids for quite a range of materials.

## 2.4 Injection of Particles into Melts

The phenomena that occur when particles are injected into melts have been extensively studied, notably by Engh and Bertheossen and Engh et al. [22], Perkins et al. [23], and Irons [24], whose recent article represents an excellent review of the current state of knowledge.

In injecting particles into melts, the main objective is to bring the particles into intimate contact with the melt and avoid the carrier gas constituting a barrier between these two phases.

Figure 2.21, taken from an article by Irons [24], shows schematically the behavior of both fine and coarse particles upon injection. It is seen that for coarse particles,

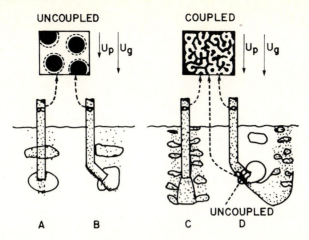

**Fig. 2.21(a).** Schematic diagram illustrating the coupling between gas and solid phases for various configurations. (A) Coarse particles injected straight downward are uncoupled so that bubbles form. (B) Coarse particles injected at an angle segregate to the bottom of the lance, and bubbling is produced. (C) Fine particles injected straight downward are coupled with the gas, and therefore form jets that penetrate until their momentum is dissipated. Bubbles with particles inside rise from this point. (D) Fine particles injected at an angle segregate to the bottom of the lance, and some gas penetrates into the bath. At the lance tip, most of the gas is in the top part of the lance and therefore the gas is uncoupled from the powder and bubbles form from the lance tip as well.

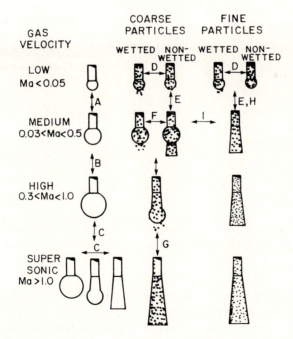

**Fig. 2.21(b).** Summary of the different possible regimes in the injection of gases and gas–solid mixtures into liquids, after Irons [24].

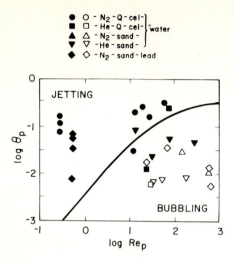

**Fig. 2.22.** Particle relative Reynolds number versus critical particle volume fraction for coupled flow. Jetting results are indicated with solid shapes, bubbling with hollow shapes, and transitions with half-shaded shapes. After Irons [24].

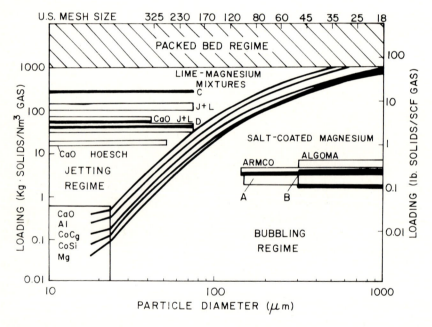

**Fig. 2.23.** Solids loading as a function of particle size for several industrial processes based on magnesium reagents. The calculated transition from bubbling to jetting is shown as well. References for each process are as follows: for salt-coated magnesium, Armco and Algoma; for Lime–magnesium mixtures, Jones and Laughlin; and for Lime, Jones and Laughlin and Hoesch. After Irons [24].

the gaseous boundary layer is only a small fraction of the particle volume, which can result in the uncoupling of the gas and solid flow, while the opposite is true for the fine particles; here, the gaseous boundary layer represents a significant fraction of the particle volume. Hence, there will be coupling between the gas and solid flow.

Figures 2.22 and 2.23 show the combined effect of particle size and particle

**Fig. 2.24.** The bath cooling rate as a function of the silica injection rate for the straight (vertical) hockey-stick and inclined lances as indicated. After Irons [24].

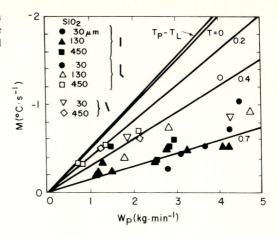

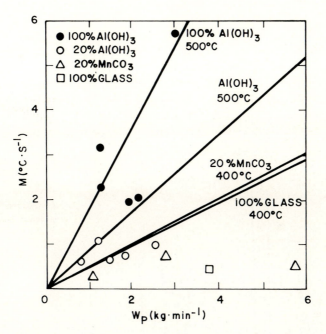

**Fig. 2.25.** Bath cooling rate as a function of solids injection rate for $Al(OH)_3$, $MnCO_3$, and glass shot mixtures. The solid lines show the cooling rates expected if all the particles were heated to the indicated bath temperature. After Irons [24].

loading on the system performance; these figures clearly indicate that small particles and high loading levels will lead to a jetting regime, while large particles and low loading levels will lead to the less desirable bubbling regime.

The fact that the injected particles may not be in ideal contact with the bath has been elegantly illustrated by Irons, who has studied the rate at which a molten metal bath was cooled by the injection of silica (see Fig. 2.24), $Al(OH)_3$ and $MnCO_3$ particles (see Fig. 2.25). It is seen that for the inert, i.e. silica, particles, the cooling rate was much less than expected on the basis of assuming that the particles equilibrated with the melt; in contrast, for the particles which did undergo thermal decomposition with significant gas evolution, the theoretical cooling rates were much more closely approached.

## 2.5 The Kinetics of Ladle Metallurgy Operations

In the previous sections, we examined the factors that govern the equilibria between phases in ladle metallurgy systems, that is, the partitioning of sulfur between the slag and the metal phases and the partitioning of oxygen between the metal–slag and gas phases.

Attainment of this equilibrium represents the theoretical limits of refining or recovery that is possible. In practice, we often fall short of these theoretical limits because of another set of factors that govern the *rate at which equilibrium is being approached*.

Figure 2.26 shows the principal regions of heat and mass transfer in ladle metallurgy systems; it is seen that we may be concerned with trasnsfer between slag and metal phases, gas–metal phases, transfer between the wall and the melt, transfer between the slag and surrounding environment, and with a three-phase gas–solid–melt region in the gas plume. In examining these systems, first we have to identify what are the rate controlling steps and then discuss how we may influence the overall process kinetics through changing the appropriate process parameters.

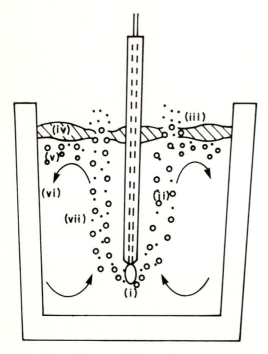

**Fig. 2.26.** The principal reaction zones during powder injection.

As a first step, let us consider a single phase system, such as that sketched in Fig. 2.27, where a fluid is made to react with a solid surface. An example of such a situation may be the rate at which a carbon lining will dissolve in steel, for example,

$$C \rightarrow \underline{C},$$

or the rate at which silica in a refractory wall would react with aluminum in the melt,

$$3SiO_2 + 4\underline{Al} = 3\underline{Si} + 2Al_2O_3,$$

and the like.

A schematic sketch of the intermediate steps involved is also shown in Fig. 2.27, which illustrates that we must consider:

1. transport of the material from the bulk to the vicinity of the reaction surface;
2. transport across a boundary layer, which may be characterized by a mass transfer coefficient; and
3. the actual chemical kinetic step.

In the majority of cases, the actual chemical kinetics will not be rate controlling, and the rate will be determined by a combination of bulk transport, that is, mixing in the bulk of the melt, and the local mass transfer. It is noted that the overall driving force for the transfer is provided by the concentration difference, that is,

$$(C_B - C_{eq}),$$

where $C_B$ is the concentration of the transferred species in the bulk and $C_{eq}$ is the equilibrium concentration at the interface; and the transfer rate is given as

$$N_A = h_D(C_B - C_{eq}), \tag{2.22}$$

where $N_A$ is the mass or molar flux and $h_D$ is the mass transfer coefficient, so that factors that govern $C_{eq}$, such as temperature and melt composition, can play an important role in affecting the overall rate.

The situation is rather more complex when we are dealing with transfer between two phases, as sketched in Fig. 2.28. These could involve gas–melt or gas–slag transfer. Here it is seen that there is a discontinuity in the concentration profiles at the interface, and the concentration of the transferred species in one phase is related to its concentration in the other, through the equilibrium relationships discussed in the first section.

Here, the rate of transfer is given as

$$N = h_{D,s}(C_{s,B} - C_{s,eq}) = h_{D,M}(C_{M,eq} - C_{M,B}). \tag{2.23}$$

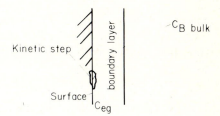

**Fig. 2.27.** Sketch of the reaction between a fluid and a solid surface.

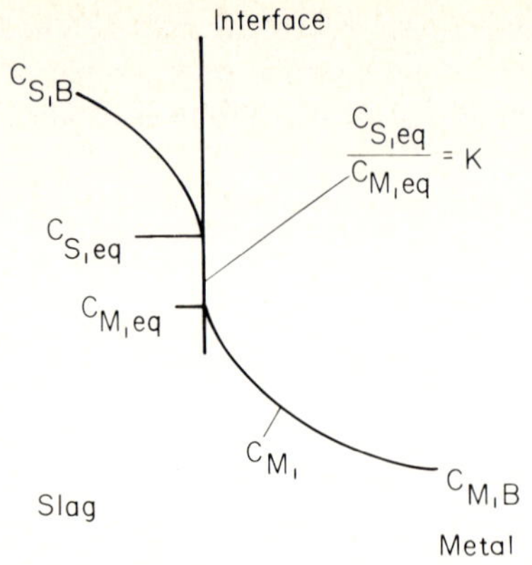

Fig. 2.28. Mass transfer between phases.

Since

$$\frac{C_{s,eq}}{C_{M,eq}} = K,$$

we have

$$N = h_{D,s}(C_{s,B} - C_{s,eq}) = h_{D,M}[KC_{s,eq} - C_{M,B}]. \tag{2.24}$$

Upon eliminating $C_{s,eq}$, we finally have

$$N = \frac{(C_{s,B} - C_{M,B})}{\dfrac{1}{h_{D,s}} + \dfrac{1}{h_{D,M}}}. \tag{2.25}$$

Here, $h_{D,s}$ and $h_{D,M}$ are the mass transfer coefficients in the slag and in the metal phases.

The very key point to stress here is that the relative importance of the mass transfer rates in the two phases will depend not only on the values of the mass transfer coefficient but also on the equilibrium constant or the partition coefficient in the system.

We shall return to the concept of the mass transfer coefficient, and in particular to the way in which the mass transfer coefficient may be predicted after we have considered fluid flow and mixing.

## 2.6 Fluid Flow and Mixing in Ladle Metallurgy Systems

Figure 2.29 shows a sketch of the velocity field and the mechanisms of mixing that one may expect in a gas-bubble stirred system. In a qualitative sense, similar considerations will apply to all metallurgical operations, in that mixing takes place by two mechanisms, namely,

1. bulk flow, that is, the macroscopic circulation that is produced by gas plumes, in this instance, or by electromagnetic forces in inductively stirred units; and

**Fig. 2.29.** Schematic sketch of the flow pattern in an argon-stirred ladle.

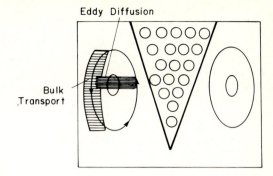

2. eddy or turbulent diffusion, which is due to the dissipation of turbulent kinetic energy in the system.

These two phenomena are of course interrelated, because turbulence is being produced because of the bath circulation. It has been shown that the principal factor that determines the rate of mixing is eddy diffusion, which in turn is related to the rate of energy dissipation in the system. This provides a fundamentally based explanation for the fact that the mixing time in most ladle metallurgy operations may be related to the rate of energy input.

Such a plot is shown in Fig. 2.30(a) [25, 26], where it is seen that argon-stirred ladles would represent the lower end of the scale, while the AOD system, with very high gas input rates, would represent the upper limit. Electromagnetic stirring would occupy an intermediate position. As we shall discuss subsequently, electromagnetically stirred systems have other rather interesting attributes that could make them attractive for a number of specialized applications. Figure 2.30(b) shows a similar plot, after Sano and Mori [27], which is somewhat more general in that allowance is also made for the vessel geometry through the $(D^2/H_o)^{1/2}$ term.

The form of the expression on the horizontal axis also indicates that it is more difficult to mix relatively shallow systems, for example, melts in electric arc furnaces or in torpedo ladles.

For argon-stirred ladles, the plot shown in Fig. 2.30(a) also agrees with the empirical expression that has been proposed for the mixing time:

$$\tau = 600\,\dot{\varepsilon}^{-0.4} \tag{2.26}$$

$$\dot{\varepsilon}(\text{W/ton}) = 14.2\frac{\dot{V}T}{M}\log(1 + \frac{H}{1.46}P_o), \tag{2.27}$$

where $\dot{\varepsilon}$ is the rate of energy dissipation; $V$ is the volumetric gas flow rate, Nm/min; $T$ is the temperature, K; $M$ is the mass of steel in tons; $H$ is the injection depth, m; and $P$ is the pressure in bar. The concept of the mixing time is very good for a preliminary assessment of the system performance; but, in many instances, much more detailed information would be desirable on the velocity fields and on the turbulent energy distribution.

Such detailed information may now be readily generated through the solution of the turbulent Navier–Stokes equations, which have the following form [28, 29, 30]:

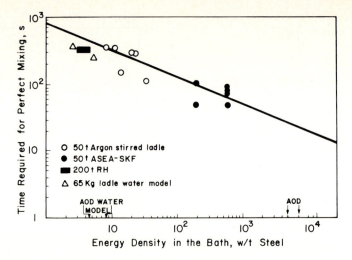

**Fig. 2.30(a).** Relationship between the time required for complete mixing and the rate of dissipation of energy per unit volume in the bath, after Nakanishi et al. Szekely [25].

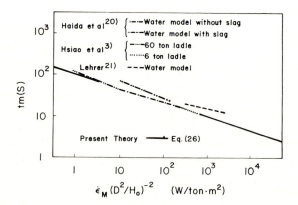

**Fig. 2.30(b).** The relationship between mixing time and the rate of energy dissipation, with a correction factor allowing for geometry, after Sano and Mori [27]. Here $D$ and $H_o$ designate the diameter of the vessel and the melt depth, respectively.

$$\nabla \cdot \rho \mathbf{U}\mathbf{U} = \nabla \mu_e \nabla \mathbf{U} + \rho \mathbf{F}_b. \tag{2.28}$$

Here, $\mathbf{U}$ is the velocity vector; $\mu_e$ is the turbulent viscosity, which has to be calculated using a turbulence model; and $\mathbf{F}_b$ is the body force.

In thermal natural convection, which will occur in all ladle metallurgy operations, but will hardly ever be dominant, this body force is due to temperature gradients, which in turn produce a density gradient. In gas-stirred systems, major density differences will exist in the plume region, which can give rise to quite strong circulation. In induction stirring, $\mathbf{F}_b$ represents an electromagnetic body force, which can provide very strong agitation. At the present time, Eq. (2.28) is readily

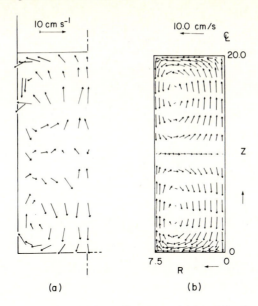

(a)        (b)

**Fig. 2.31.** Comparison between the experimentally measured (a) and the theoretically predicted (b) velocity profiles in a laboratory-scale induction furnace [31].

**Fig. 2.32.** Comparison of the experimentally measured velocity as a function of the coil current for a four-ton inductively stirred molten steel bath [32].

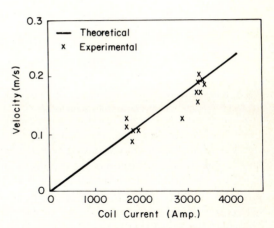

solved using digital computers, and possibly appropriate computational packages. Figures 2.31(a) and (b) show a comparison between experimentally measured and theoretically predicted circulation patterns in a laboratory-scale inductively stirred melt [31], while Fig. 2.32 shows a comparison between experimentally measured and theoretically predicted melt velocities in a four-ton steel melt.

Figure 2.33 shows a comparison between measurements and predictions for a water model of an argon-stirred ladle; here again, there is good agreement between the computed and experimentally obtained data.

On this basis, we can proceed to compute the velocity profiles and the turbulent characteristics of a range of industrial systems, and a selection of these data is shown in Figs. 2.34–2.36.

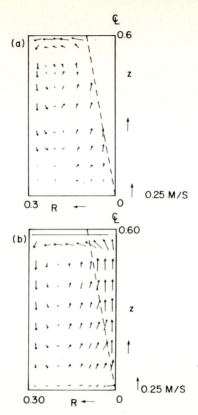

**Fig. 2.33.** A comparison between the (a) experimentally measured and (b) theoretically predicted velocity profiles in a water model of an argon-stirred ladle.

Figure 2.34 shows the computed velocity fields and the maps of the turbulent kinetic energy for an argon-stirred ladle [32]; the corresponding values for an induction-stirred vessel are shown in Fig. 2.35, and those for an R–H degasser are given in Fig. 2.36 [33]. Inspection of these plots shows the following general behavior.

For argon-stirred ladles, the melt velocity is high in the plume and at the free surface, but there are significantly quiescent regions at the bottom, especially in the vicinity of the wall. The turbulent kinetic energy follows a similar pattern. For an inductively stirred vessel, the maximum melt velocities tend to be higher than for gas stirring; at the same time, both the velocity fields and the maps of the turbulent kinetic energy are much more uniform. In a properly designed, inductively stirred system, there are no stagnant regions. The velocity maps and maps of the turbulent kinetic energy in R–H degassers are an intermediate between the electro-magnetically and the gas-bubble stirred systems.

The fluid flow behavior of these systems provides very useful guidance regarding their potential application in ladle treatment. If uniformity of the velocity field and of the turbulent kinetic energy is important and very high turbulence levels at the

**Fig. 2.34(a).** The computed velocity field for a six-ton ladle holding molten steel, agitated by an argon gas stream.

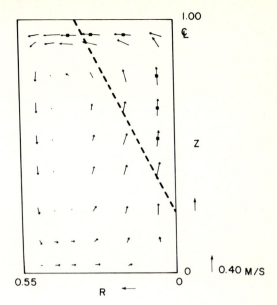

**Fig. 2.34(b).** The computed map of the turbulent kinetic energy for a six-ton ladle holding molten steel, agitated by an argon gas stream.

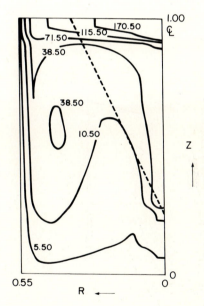

free surface are not required, inductive stirring is the best choice. If promoting slag–metal reactions is important and one can tolerate relatively stagnant regions in the vicinity of the bottom of the ladle, gas stirring would be ideal—from a process standpoint.

As far as agitation is concerned, an R–H degassing system would represent an intermediate between these extremes; an R–H system would of course be preferred

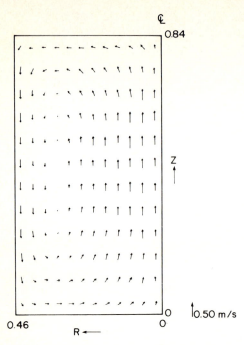

Fig. 2.35(a). The computed velocity field for a four-ton inductively stirred steel melt.

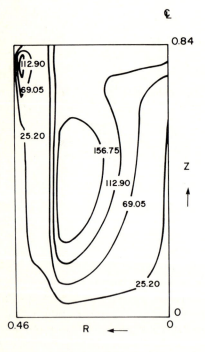

Fig. 2.35(b). The computed map of the turbulent kinetic energy for a four-ton inductively stirred steel melt.

**Fig. 2.36.** The computed velocity field for an industrial-scale R–H vacuum degassing unit, after Shirabe and Szekely [33].

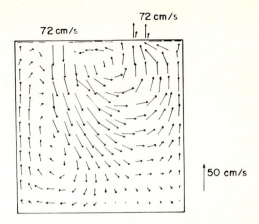

if actual degassing is sought, because the imposition of the vacuum would promote degassing from the standpoint of thermodynamic equilibrium considerations.

### 2.6.1 Flotation of Inclusions

In a quiescent melt, the rate at which inclusion particles would float may be calculated by equating the buoyancy forces with the viscous forces to obtain Stokes's Law, which is written as

$$U_t = \frac{d_p^2}{18\mu}(\rho_p - \rho_f)g, \tag{2.29}$$

where $U_t$ is the terminal rising velocity, $d_p$ is the particle diameter, $\rho_f$ is the density of the melt, $\rho_p$ is the density of the particle, and $\mu$ is the viscosity of the melt.

Upon substituting appropriate numerical values, the rising velocity for a small, for example, 20-$\mu$m alumina particle in steel would be

$$1.7 \times 10^{-4} \text{m/s},$$

while the corresponding value for a 100-$\mu$m particle is

$$4.2 \times 10^{-3} \text{m/s}.$$

These rising velocities are quite slow, and for a quiescent melt, only the quite large particles would be able to float out during the treatment time, which has to be of the order of a few minutes.

When the melt is in motion, the sitation is somewhat more complicated. Highly turbulent conditions will promote the collision of the inclusion particles, and hence their agglomeration. The larger the particles, the higher is their rising velocity, as per Eq. (2.29).

At the same time, when the flow is highly turbulent, the fluctuating velocity component in the turbulent flow may markedly influence particle motion. In particular, when the fluctuating velocity component becomes comparable to the rising velocity of the particle, particle flotation may be severely inhibited. For this reason, the best arrangement may be to stir vigorously at first, in order to promote agglomeration, and then to stir very gently; "rinsing" is the term usually employed to promote the flotation of the inclusion particles.

## 2.7 Convective Mass Transfer and Kinetics

Once the velocity field and the maps of the turbulent kinetic energy are known, we can readily address the question of mass transfer problems, such as the rate of dispersion of alloying additions, desulfurization, or dephosphorization kinetics [34].

The conservation of a given transferred species is given by the expression

$$\frac{\partial C_A}{\partial t} + \mathbf{U} \nabla C_A = \nabla D_{\text{eff}} \nabla C_A + \dot{r}_A, \qquad (2.30)$$

where $C_A$ is the concentration of the transferred species, such as an alloying element, sulfur, or oxygen; $\dot{r}_A$ is the rate of chemical reaction; and $D_{\text{eff}}$ is the eddy diffusivity, which is related to the eddy viscosity by

$$\frac{\mu_t}{D^t \rho} \approx 1. \qquad (2.31)$$

Here, $\mu_t$ and $D^t$ are the turbulent viscosity and the turbulent diffusivity, respectively. In order to complete the statement of the problem, we must state the boundary conditions, which in the present case will have to specify some rate of reaction or

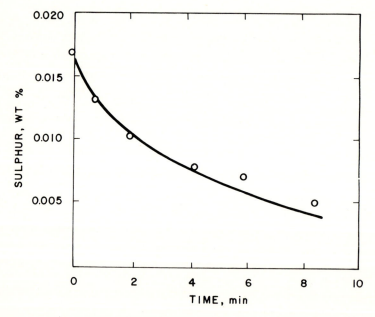

**Fig. 2.37.** A comparison of the experimentally measured (data points) and the theoretically predicted sulfur content during the desulfurization of a 40-ton vessel holding molten steel. [29].

equilibrium condition at the slag–metal interface or at the plume–melt interface. Numerous calculations of this type have been published in the literature [35, 36], and Figs. 2.37–2.39 give some examples concerning desulfurization kinetics.

An alternative, somewhat simplified approach, but one which can provide useful insight, is to consider the whole melt well stirred and assume that all the resistance to transfer is concentrated at the phase boundary, that is, at the melt–slag or melt–plume interface.

Then we can write that the molar flux to the slag–metal interface is given by

$$N = h_D(C_B - C_{eq}),\tag{2.32}$$

where $h_D$ is the previously defined mass transfer coefficient.

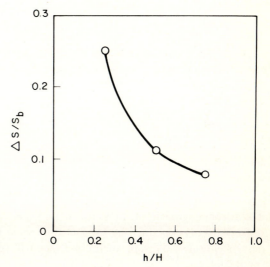

**Fig. 2.38.** Sulfur segregation in a molten steel bath undergoing desulfurization as a function of the lance depth [29].

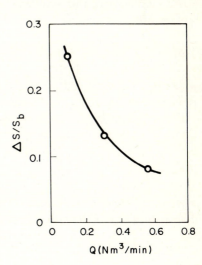

**Fig. 2.39.** Sulfur segregation in a steel bath undergoing desulfurization as a function of the carrier gas flow rate [29].

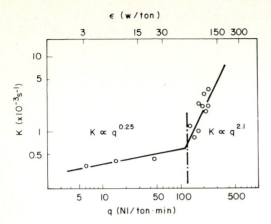

**Fig. 2.40.** The effect of the stirring gas flow rate on the mass transfer capacity coefficient, $K$ characterizing the desulfurization reaction, after Asai et al. [36]. Also shown is the rate of turbulent energy dissipation in the bath. The relationship between $K$ and $h_D$ is discussed in [36].

Numerous correlations have been proposed to predict $h_D$ as a function of the operating conditions. As discussed in an elegant paper by Asai [36], and shown in Fig. 2.40, these seem to be consistent regarding the fact that the mass transfer coefficient is a function of the turbulent kinetic energy dissipation; indeed, for low values of $\dot{\varepsilon}$, for example, $\dot{\varepsilon} < 60$ W/ton,

$$h_D \alpha \dot{\varepsilon}^{1/4}. \tag{2.33}$$

For high values of $\dot{\varepsilon}$, that is, $\dot{\varepsilon} > 60$ W/ton, we have

$$h_D \alpha \dot{\varepsilon}^{2.1}. \tag{2.34}$$

The abrupt change in the mass transfer coefficient at dissipation values in the range of 60 W/ton is explained by the fact that when this energy input rate is exceeded, emulsification of the slag–metal interface will take place, with a greatly increased interfacial area.

We may address this problem in a somewhat different manner by defining a modified Biot number as

$$N'_{\text{Bi}} = \frac{h_D L}{D_{\text{eff}}}, \tag{2.35}$$

where $D_{\text{eff}}$ is the eddy diffusivity in the melt, determined by the rate of agitation; $L$ is the melt depth; and $h_D$ is the mass transfer coefficient.

For situations where slag emulsification does not take place, $h_D$ may be estimated from a Higbie-type relationship of the following type:

$$h_D \sim \sqrt{\frac{D}{\pi t_e}}; \qquad t_e \cong \left| \frac{R}{U_c} \right|, \tag{2.36}$$

where $U_c$ is a characteristic melt velocity, $R$ is the crucible radius, $D$ is the molecular diffusivity, and $t_e$ is the exposure time.

For typical argon-stirred ladles, $U \sim 30$ cm/s, $R = 1.5$ m, $H = 3$ m, and $D = 5 \times 10^{-5}$ cm/s.

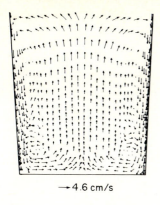

→ 4.6 cm/s                                     (a) The velocity field after 5 min.

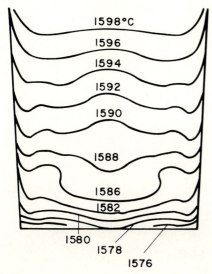

1598°C

1596

1594

1592

1590

1588

1586

1582

1580

1578

1576                                           (b) The temperature field after 5 min.

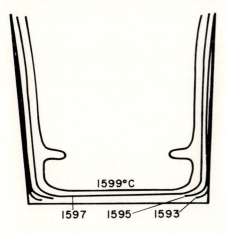

1599°C

1597      1595      1593                       (c) The temperature field after induction stirring.

**Fig. 2.41.** Calculations concerning the stratification of an 80-ton ladle holding molten steel.

$$N_{Bi} = \frac{3.2 \times 10^{-3} \times 150}{10} \sim 5 \times 10^{-2}$$

It follows that when there is no emulsification, the modified Biot number is small; hence, the overall rate will be limited by mass transfer at the slag–metal interface. In contrast, when emulsification occurs, the effective mass transfer coefficient will increase by several orders of magnitude and at that time the rate-limiting step will be eddy diffusion from the bulk to the interface.

## 2.8 Heat Transfer

Heat transfer is an important component of all ladle metallurgy and injection systems. The following key points have to be considered here.

1. When molten steel is held in a ladle, heat losses will occur to the surroundings, and in particular to the ladle walls. In addition to a modest rate of heat loss, this will also result in stratification of the melt.
2. When molten steel is being agitated in a ladle by gas injection or induction stiring, the rate of heat loss will be increased, particularly because of radiative heat transfer from the free surface.
3. In many ladle metallurgy operations, thermal energy is added to the melt either by induction or, more frequently, through the use of electric arcs. The operation of these "ladle furnaces" is of considerable and growing interest.

Fundamental consideration can provide us with useful guidance regarding all three of these aspects of injection metallurgy heat transfer.

### 2.8.1 Heat Loss from Quiescent Melts in Ladles

When molten steel is held in a ladle in the absence of external agitation, heat will be lost both to the refractory walls by conduction and from the free surface by thermal radiation [37, 38]. If, as a first approximation, we assume that there are no large temperature differences in the melt (in fact, these temperature differences will usually be less than about 20–30°C), then we may estimate the conductive heat loss into the lining by solving the unsteady state heat conduction equation [39], which will give us this relationship:

$$Q = 2kA(T_0 - T_i)\left(\frac{t_e}{\pi\alpha}\right)^{1/2},$$

where $Q$ is the rate of heat transfer (in joules), $T_0$ is the steel temperature, $T_i$ is the initial refractory temperature, $A$ is the total refractory area, $\alpha$ is the thermal diffusivity of the refractory, $k$ is the thermal conductivity of the refractory, and $t_e$ is the exposure time. Then the rate of heat loss is given as

$$mC_p(T_{i,m} - T_{f,m}) = Q, \tag{2.37}$$

where $m$ is the mass of steel in the ladle, $C_p$ is the specific heat of the steel (700 J/kg K), $T_{i,m}$ is the initial steel temperature, and $T_{f,m}$ is the final steel temperature. For typical steel processing conditions, this expression would give heat loss rates of the order of 0.5–1°C/min.

The radiative heat loss from the steel surface would be given by the following expression:

$$\dot{q}'' = \varepsilon \sigma T_s^4, \tag{2.38}$$

where $\varepsilon$ is the emissivity of the metal surface, $\dot{q}''$ is the heat flux in W/m$^2$, $\sigma$ is Boltzmann's constant $= 5.67 \times 10^{-8}$ W/m$^2$K$^4$, and $T_s$ is the surface temperature. For free, exposed metal surfaces, this would give a very high rate of heat loss, however, it has been shown that a slag cover of, for example, 5–15 cm thick can minimize the free surface heat loss. However, when the slag cover is disrupted, for example, by gas bubbling, significant incremental heat loss may occur through the top surface.

It has been shown that heat loss by conduction through the side walls can set up thermal natural convection in the melt, which will provide some measure of mixing. This effect can be readily modeled, and Figs. 41(a), (b), and (c) show the computed velocity and temperature profiles for an 80-ton ladle; the circulation pattern, downward near the side walls and upward in the middle, is typical of what one would expect. It is seen, furthermore, that significant, that is, for example, a 20°C, stratification can take place here. However, this stratification may be readily minimized by quite gentle agitation, either by bubbling or by induction stirring.

## 2.8.2 Heat Loss from Agitated Melts

In typical ladle treatment operations, the melt is agitated by the injection of gases and solid particles. Under these conditions, the heat loss from the system will increase because of the disturbance to the slag cover, and also because of endothermic reactions and rapid gas evolution (in case of carbonate injection). The precise prediction of this enhanced rate of heat loss is quite difficult, but Table 2.5, after Szekely and Lehner [38], may provide a useful guide.

More about the practical aspects of heat losses due to ladle treatment is discussed in Chapter 3.

**Table 2.5.** Typical heat loss values in ladle treatment

| Treatment | Value[a] |
|---|---|
| Holding, in the absence of agitation | 1°C/min |
| Gas bubbling | 2°C/min |
| Solid injection | 3–3.5°C/min |

[a] These figures would be somewhat higher for smaller ladles.

### 2.8.3 Ladle Furnace Operations

One of the important new developments in ladle metallurgy is the possibility of adding thermal energy to the melt by induction or the use of arcs. The operation of such "ladle furnaces" is attractive, because one may compensate for the heat losses in ladle treatment as part of the treatment process, rather than requiring the expensive incremental furnace holding time in the melting or steelmaking part of the process.

One key question in the design of ladle furnace systems is whether the thermal energy supplied by induction or by arcs can be readily dissipated in the ladle, or more specifically, the extent of agitation needed to bring about thermal uniformity. It is an established fact that in induction furnace operations, heat transfer and agitation occur simultaneously, so one should not expect any thermal gradients.

The situation is a little more complex in the case of arc heating, because here the thermal energy is supplied at, or near, the free surface, which has to be dispersed by agitation. Indeed, some form of agitation, either by induction or by gas bubbling, must be an inherent feature of any ladle furnace system.

In order to assess the extent of agitation needed, let us define a ladle furnace Biot number in the following manner:

$$N_{\text{Bi}} = \frac{hL}{k_{\text{eff}}}, \tag{2.39}$$

where $h$ is the heat transfer coefficient between the melt and the arc, $k_{\text{eff}}$ is the thermal conductivity of the melt, and $L$ is the characteristic length, which is the melt depth. When the Biot number is large, surface heat transfer is limiting and then we can expect significant thermal gradients; in contrast, when the Biot number is small, internal conduction or convection, represented by $k_{\text{eff}}$, is limiting.

Let us consider a typical ladle dimension of 3 m and the fact that typical values of the free surface heat transfer coefficient should be in the 2000–5000 W/m² K range. On noting that the atomic thermal conductivity of steel is about 30 W/m K, in the absence of agitation, the Biot number would be of the order of

$$\frac{2000 \times 3}{20} \sim 300,$$

that is, very large, resulting in significant temperature gradients. However, when the system is agitated, we have to work in terms of the turbulent properties.

One may state that the turbulent Prandtl number is unity, that is,

$$N_{\text{Pr}}^{(t)} \cong \frac{C_p \mu_t}{k_t} \approx 1. \tag{2.40}$$

Since the turbulent viscosity in most agitated systems is some 100–500 times the atomic value, we may write

$$\mu_t \sim 300 \times 6.7 \times 10^{-3} \sim 2.0 \text{ kg/m s}.$$

Since $C_p \sim 700$ J/kg K,

$$k_{\text{eff}} \approx k_t \sim 1{,}400 \text{ W/m K},$$

which would give a Biot number of the order of

$$\frac{2,000 \times 3}{1,400} \sim 5.$$

This implies that on arc heating of a melt, there will be temperature gradients in the system, but that these would not be very large; furthermore, once the heat supply is discontinued, the melt could be quite rapidly homogenized. It follows that from the standpoint of dispersing temperature inhomogeneities, quiet, gentle agitation should be quite adequate, so that either gas-bubble stirring or induction stirring should do the job.

# References

1   L.S. Darken and R.W. Gurry, "Physical Chemistry of Metals," McGraw-Hill, New York (1953)
2   Y.K. Rao, "Stoichiometry and Thermodynamics of Metalllurgical Processes," Cambridge Univ. Press, London and New York (1985)
3   D.R. Gaskell, "Introduction to Metallurgical Thermodynamics," 2nd ed., McGraw-Hill, New York (1981)
4   E.T. Turkdogan, "Physical Chemistry of High-Temperature Technology," Academic Press, San Diego (1980)
5   E.T. Turkdogan, "Physicochemical Properties of Molten Slags and Glasses," Metals Society, London: (1983)
6   R.J. Fruehan, "Ladle Metallurgy: Principles and Applications," ISS-AIME, Warrendale, PA (1985)
7   I.D. Sommerville, "The measurement, prediction and use of capacities of metallurgical slags," Proc. SCANINJECT IV, Lulea, Sweden: MEFOS (1986)
8   T. Emi and Y. Iida, "Impact of injection metallurgy on the quality of steel products," Proc. SCANINJECT III, Lulea, Sweden: MEFOS (1983)
9   E.T. Turkdogan, "Technology innovations in pneumatic steelmaking and ladle-refining processes during the 1980's," Proc. SCANINJECT IV, Lulea, Sweden: MEFOS (1986)
10  E.T. Turkdogan, "Physicochemical properties of molten slags and glasses," The Institute of Metals, London (1983)
11  E.T. Turkdogan, in "Perspective in metallurgical development 1884–1984," The Institute of Metals, London, pp. 49–60 (1984)
12  B. Bjorkman and E. Jacobson, "New thermodynamic tools," Proc. SCANINJECT IV, Lulea, Sweden: MEFOS (1986)
13  M. Sano and K. Mori, "Circulating flow model 'in a molten metal bath with special respect to behaviour of bubble swarms and its application to gas injection procedures," Proc. SCANINJECT III, Lulea, Sweden: MEFOS (1983)
14  J. Szekely, "Fluid Flow in Metals Processing," Chapter 8, Academic Press, San Diego (1979)
15  G. Carlsson and E. Burstrom, "Alternative tuyere design: prolonging the life of the tuyere," Proc. SCANINJECT IV, Lulea, Sweden: MEFOS (1986)
16  Y. Ozawa and K. Mori, Trans. I.S.I.J. 26. (4), 291 (1986)
17  J. Szekely, "Fluid Flow in Metals Processing, Chapter 7, Academic Press, San Diego (1979)
18  J.F. Davidson and D. Harrison, "Fluidization," Elsevier, Amsterdam (1973)
19  J. Szekely, "Fluid Flow in Metals Processing," Chapter 7, Academic Press, San Diego (1979)
20  D. Kunii and D. Levenspiel, "Fluidization Engineering," Wiley, New York (1968)
21  F.A. Zenz and D.F. Othmer, "Fluidization and Fluid-Particle Systems," Van Nostrand-Reinhold, Princeton, NJ (1960)
22  (a) T.A. Engh and H. Bertheussen, Scan. J. Met. 4, pp. 241–249 (1975) (b) T.A. Engh, K. Larsen, and K. Venas, Ironmaking and Steelmaking 6, pp. 268–273 (1979)

23   A. Perkins, T. Robertson, and D. Smith, "Improvements to liquid steel temperature control in the ladle and tundish," Proc. SCANINJECT IV, Lulea, Sweden: MEFOS (1986)

24   G.A. Irons, "Fundamental and practical aspects of lance design for powder injection processes," Proc. SCANINJECT IV, Lulea, Sweden: MEFOS (1986)

25   K. Nakanishi, T. Fujii, and J. Szekely, "Possible relationships between energy dissipation and agitation in steel processing operations," Ironmaking and Steelmaking *3*, p. 193 (1975)

26   A. Murthy and J. Szekely, "Some fundamental aspects of mixing in metallurgical reaction systems," Met. Trans. *17B*, p. 487 (1986)

27   M. Sano and K. Mori, "Circulating flow model in a molten metal bath with special respect to behavior of bubble swarms and its application to gas injection processes," Proc. SCANINJECT III, Lulea, Sweden: MEFOS (1983)

28   J. Szekely, "Fluid Flow in Metals Processing," Chapter 4, Academic Press, San Diego (1979)

29   N. El-Kaddah and J. Szekely, "Desulfurization kinetics in ladle metallurgy," Proc. 3rd Process Technology Conference on Mathematical Modelling, Pittsburgh, PA, p. 221 (1982)

30   J. Szekely and N. El-Kaddah, "Turbulence phenomena and agitation in ladle metallurgy," Iron and Steelmaker *11*(1), p. 22 (1984)

31   J.-L. Meyer, N. El-Kaddah, J. Szekely, C. Vives, and R. Ricou, "A comprehensive study of the induced current, the electromagnetic force field, and the velocity field in a complex electromagnetically driven flow system," Met. Trans. *18B*, pp. 529–538 (1987)

32   N. El-Kaddah, J. Szekely, and G. Carlsson, "Fluid flow and mass transfer in a four-ton inductively stirred melt," Met. Trans. *15B*, p. 633 (1984)

33   K. Shirabe and J. Szekely, "A mathematical model of fluid flow and inclusion coalescence in the R–H vacuum degassing system," Trans. I.S.I.J. *23*, p. 465 (1983)

34   O.J. Ilegbusi and J. Szekely, "Mathematical modelling of the electromagnetic stirring of molten metal–solid suspensions, Trans. I.S.I.J., in press (1988)

35   J. Szekely and N. El-Kaddah, "Desulfurization kinetics in argon-stirred ladles," Ironmaking and Steelmaking *8*, p. 269 (1981)

36   S. Asai, M. Kawachi, and I. Muchi, "Mass transfer rate in ladle refining processes," SCANINJECT III, Lulea, Sweden: MEFOS (1983)

37   O.J. Ilegbusi and J. Szekely, "Melt stratification in ladles," Trans. I.S.I.J. *27*(7), pp. 563–572 (1987)

38   J. Szekely and T. Lehner, Notes for a short course held June 1986 at MEFOS, Lulea, Sweden (1986)

39   J. Szekely and N.J. Themelis, "Rate Phenomena in Process Metallurgy," Wiley, New York (1971)

# 3 Injection Practice in the Secondary Metallurgy of Steel

## Lars Helle

## 3.1 Introduction

In this chapter, we shall describe the practical aspects of ladle metallurgy, with emphasis on equipment construction, operating experience, and practical operating guidelines. In an ideal world, practice would be fully based on theory, and all the practices would have a complete fundamental explanation and justification. This would then represent a complete harmony between the material in this chapter and that given in Chapter 2.

Alas, this ideal has not been fully attained. Many aspects of injection practice have only partial theoretical fundamental support. Indeed, in certain cases, one has to rely on empirical or semiempirical relationships as practical guidelines. Thus, this chapter represents a somewhat different look at the same phenomena that were covered previously. These two viewpoints are often consistent but at times diverge, giving a faithfull impression of a field that is still new and rapidly developing.

## 3.2 Description of the Process

### 3.2.1 Equipment

#### 3.2.1.1 General Layout

The powder injection system as a whole is normally composed of the parts shown in Fig. 3.1 [1]. Disregarding the material handling and with the assumption that compressed carrier gas (either nitrogen or argon) is available, the equipment necessary to convey the powder pneumatically into steel includes the powder dispenser, the feeding line, the lance and lance stand or a nozzle in the ladle wall or in the ladle bottom.

The layouts of three powder injection systems from different manufacturers are shown in Figs. 3.2 [2], 3.3 [3], and 3.4 [4]. The characteristic feature of the German TN equipment (Fig. 3.2) is that the dispenser and the lance form an integrated unit capable of moving up and down a vertical stand. Other systems, for example, from Scandinavian Lancers (Fig. 3.3), have a separate dispenser that is connected by a flexible hose to the lance. The same principle has been adopted by other suppliers like OVAKO STEEL [5] in Finland, IRSID [6] in France, and Klöckner [7] and others in Germany. Yet another modification of the layout of the equipment is that of Swedish ASEA (Fig. 3.4). In this system, the dispenser and the lance form one compact unit. This type of approach shortens the distance over which the powder is to be transported to a minimum, thereby possibly reducing the need for carrier

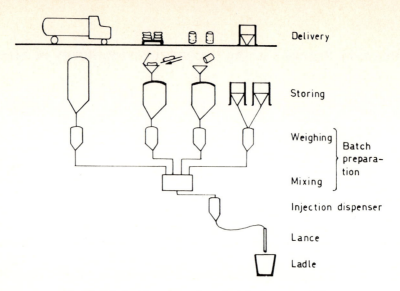

**Fig. 3.1.** Principal flow chart of powder injection system [1].

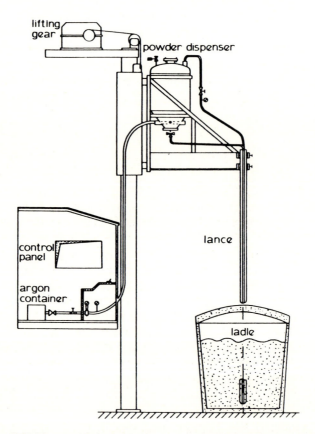

**Fig. 3.2.** Layout of TN (Thyssen–Niederrhein) injection equipment with moving powder dispenser unit [2].

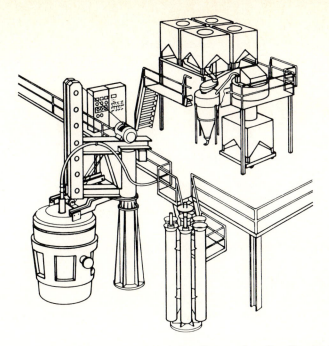

**Fig. 3.3.** Multicontainer injection system with stationary dispenser from Scandinavian Lancers [3].

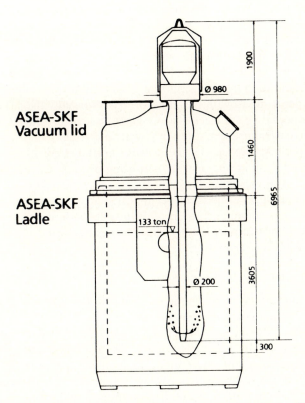

**Fig. 3.4.** ABB Metallurgy (former ASEA Metallurgy) powder injection equipment [4]. Reprinted with permission from G. Grimfjärd, ASEA-SKE, Users Seminar (1983).

**Table 3.1.** Injection System Suppliers

| Supplier | Location |
|---|---|
| ABB Metallurgy (ASEA) | Sweden |
| Clesid/IRSID | France |
| Goffart | France |
| Klöckner–Max Peter | West Germany |
| Luftfiltrering A/S | Norway |
| OVAKO STEEL Oy Ab | Finland |
| Paulus | West Germany |
| Polysius | West Germany |
| Rossborough | United States |
| Scandinavian Lancers | Sweden |
| Thyssen–Niederrhein (TN) | West Germany |
| Vulcan | United States |

gas. In operation, when used to inject powder into the ASEA–SKF ladle through a vacuum lid, it gives an airtight seal against the lid, keeping air efficiently away from the reaction site. Potential suppliers of injection systems are listed in Table 3.1.

### 3.2.1.2 Dispenser

The dispenser is basically a pressure vessel with a conical bottom. A selection of alternative designs is presented in Fig. 3.5 [8]. In modern injection stations, a storage container is located on top of the dispenser to charge the powder to be injected into the dispenser through an inlet valve. At the bottom of the dispenser, there is a metering nozzle and an outlet valve through which the powder is transported after fluidization into the ejector and the transport hose. In the ejector, the fluidized powder is mixed with the transport gas. Normally the dispenser rests on load cells connected to an electronic weighing system in order to facilitate the injection of predetermined quantities of powder material.

To be properly transportable, the powder must behave like a fluid. This is accomplished by a "fluidizer," usually consisting of small gas nozzles protruding through a false wall located in the conical part of the dispenser. By "aeration" of the powder, liquid-like properties are imported to it. A proper fluidization is essential to obtain a smooth transport of the powder without pressure fluctuations and blockage in the transport tube system. The basic concepts of fluidization have been discussed in Chapter 2.

### 3.2.1.3 Feeding Line

The fluidized powder has to be kept in this state all the way down from the ejector to the nozzle through which it enters the steel.

Downstream of the ejector the gas–solid flow can behave in different ways depending mainly on the linear gas velocity. The mass flow ratio is chosen to be within the dispersed (dilute) flow mode to minimize pressure fluctuations. To maintain the particles suspended, the average gas velocity has to be larger than the settling velocity of the solid particles. Such settling would, of course, result in the

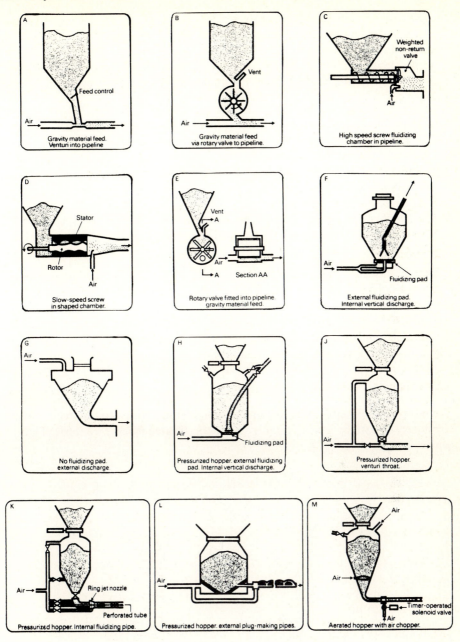

**Fig. 3.5.** Various alternatives for powder dispensers [8].

blockage of the conduit. These problems have been discussed in Chapter 2 and further detailed summaries of pneumatic conveying for feeding line design can be found in the literature [9].

As the design parameter is the gas velocity, care has to be taken to keep the pressure drop in the feeding line at a minimum. This can be achieved by the following:

1. A proper inner diameter of the tube. Employing feeding rates of, for example, 10–30 kg/min, a diameter of 19 mm has shown to be adequate.
2. A smooth inner surface of the tube.
3. A minimum number of area contractions. A maximum allowed contraction of 20% can be used as a rule of thumb.
4. A minimum number and proper radius (> 500 mm) of bends.
5. The length of transportation of the powder being as short as possible; ABB Metallurgy (ASEA) and TN have adopted this principle in their systems (see Figs. 3.2 and 3.4).

Steel pipes with smooth inner surfaces should be used in the feeding lines where ever possible, particularly in the parts subject to radiation from the molten steel surface. Reinforced plastic hose can be used in the moving parts. In such a case, they have to be manufactured to tolerate pressures up to 16 bar. In the connections, for example, between the lance and the feed line, safety couplings have to be employed of the type such as OPW Kamlock.

### 3.2.1.4 Lances

The most common apparatus to convey powdered material into the steel or affect stirring of the bath by gas injection is a refractory coated lance. Two types of lances can be distinguished:

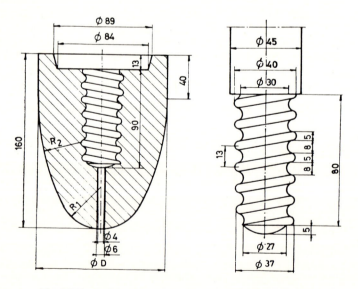

**Fig. 3.6.** Example of the construction of the lance tip [10].

1. lances made of refractory sleeves with a separate lance tip, and
2. lances made of monolithic castable refractory material with or without a separate tip.

### 3.2.1.4.1 Sleeve Lances

The most usual coating of the lance tube is a burned refractory sleeve. The lance tip in this case is a separate exchangeable part made of, for example, alumina, chamotte, or graphitized chamotte material. The seams between the sleeves are sealed by refractory cement or ceramic fiber.

The lance tube, made of steel with heavy wall thickness, has normally a threaded point to allow the lance tip to be changed as needed, for example, after failure in service. A typical example is shown in Fig. 3.6 [10].

The sleeves can be prestressed, as shown in Fig. 3.7. The prestressing is carried

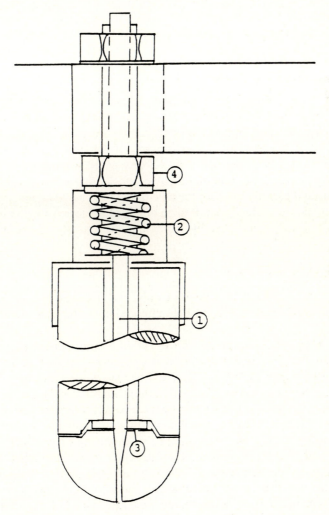

**Fig. 3.7.** Arrangement for prestressing the lance sleeves: 1, steel tube; 2, spring; 3, washer; and 4, nut [11].

out in order to maintain tight seams between the sleeves to prevent penetration by the molten steel during service. Should a sleeve of a prestressed construction fail during injection, it may, however, remain in its place because of the stresses exerted upon it.

The normal refractory material of the sleeves is chamotte (possibly a high-alumina slag line), which is preferred because of its low cost. These types of lances, with a single nozzle, have a typical service life of 10–15 min or 1–5 charges.

### 3.2.1.4.2 Monolithic Lances

Monolithic lances are manufactured of high-purity castable ceramics, for example, alumina or magnesite. Figure 3.8 [12] is an expanded view showing typical features of a monolithic lance. The reinforcing of the monolithic lances intended for multiple use is a decisive factor for their service life. Various alternatives have been tried, for example, reinforcing bars, angle iron, chicken wire, barbed wire, and steel fibers. The service life of monolithic lances is around 100 mins. Results for cast magnesite lances, with intermediate repair, of 50 charges or more, have been reported [7].

### 3.2.1.4.3 Prevention of Nozzle Blockage

The lance outlet nozzle is a critical point in the transportation system. It must be correctly dimensioned in relationship to the gas flow rate, powder feeding rate, and lance immersion depth, in order to keep the nozzle open.

It has been established that the penetration of the melt into the lance nozzle during injection is diminished when the velocity of the gas through the nozzle is increased. In practice, the minimum tolerable velocity depends on the powder flow rate and is around 80 and 25 m/s for the flow rates of 10 and 30 kg/min, respectively. The minimum desired velocity can best be obtained by decreasing the surface area of the nozzle opening and thereby increasing the velocity of the gas–powder suspension.

### 3.2.1.4.4 Miscellaneous Aspects

The lances can be constructed to have only one opening or several openings (e.g., the multinozzle lances), the most common number of nozzles being three (see Fig. 3.8).

The advantage of the multinozzle lances is that they stabilize the bubble plume ascending in the bath, thereby reducing surface turbulence, splashing, and ejections. They give greater mixing efficiency [13] and increased mass transfer rates [14] as well. Further, it has been reported [134] that the angled lances achieve better liquid–particle contact because the particles and gas are largely separated before they leave the lance (see Chapter 2 for more details). Arguably similar considerations may apply to multihole lances, where the nozzles are at an angle to the lance tube.

An additional advantage of the multinozzle lances is that they assist in reducing the consumption of the carrier gas. At OVAKO STEEL, Imatra Steelworks' introduction of multinozzle lances helped to reduce the carrier gas consumption by 50% with a simultaneous increase in the powder flow rate from 25 to 30 kg/min.

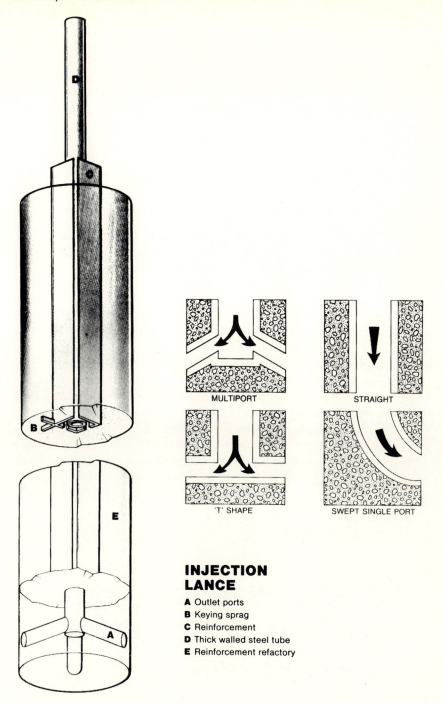

**INJECTION LANCE**

**A** Outlet ports
**B** Keying sprag
**C** Reinforcement
**D** Thick walled steel tube
**E** Reinforcement refactory

**Fig. 3.8.** An example of a monolithic lance together with alternative multiport designs [12].

### 3.2.1.5 Injection Through a Slide Gate

Stopinc Aktiengesellschaft has developed a method to apply a two-hole slide gate system for injecting solids. The idea is to replace one of the two collector nozzles by an injector and to carry out the metallurgical treatment directly through the existing nozzle of the slide gate instead of the lance.

Fig. 3.9 [15] shows a two-hole slide gate consisting of the following parts, known from the linear slide gate:

frame–housing–cover–pushrod
cassette
slider with 2 nozzles
exchangeable collector nozzle
injector

The same refractories are used as for the standard linear slide gate (i.e., well block, ladle nozzle, cassette plate, and slider plate with 2 holes). Instead of the linear slide gate, a revolving gate could also be used for the process described.

In connection with conventional casting, that is, open top or bottom pouring, the application of a two-hole slide gate for injection is relatively easy to realize on ladles up to approximately 100 tons. It may become necessary to open the slide gate by oxygen lancing, since the filler sand has been drained before injection. As a consequence, steel will freeze in the ladle nozzle between the end of treatment and the beginning of pouring.

The use of slide gate injection may be difficult in continuous casting, especially with sequence casting, where a self-opening of the gate is important because of short time availability and the limited space between tundish and ladle. To overcome this problem, a separate one-hole slide gate for injection only could be a solution, provided the gate refractories would last the life of the ladle bottom lining. This might be feasible as there would be no pouring action through such a nozzle. Trials with this system have been carried out at two steel plants in Germany [137].

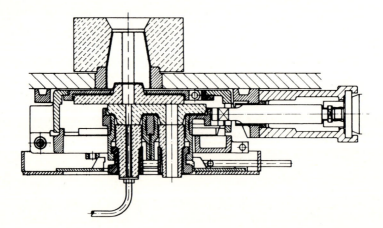

**Fig. 3.9.** Slide gate to be used for injecting powder material into steel [15].

Although the information is scarce some success has apparently been achieved. The viability of this system in full production conditions in day-to-day operation has still to be proven.

### 3.2.1.6 Injection Through the Ladle Wall

The Injectall Limited of Sheffield, U.K. [142], has developed a ladle side injection device named ISID. This device is located in the side of the ladle, near the bottom, to create a good, intense stirring effect. According to the company, the mechanism is designed to have a metal-tight system for the injection of gases, powders, and wires with a safe reliable shutoff after use.

The ISID mechanism is actuated hydraulically for remote operation. The system advances a lance to punch out a refractory sealing plug and also employs a separate shut-off device at the end of the tratment to make the operation absolutely safe and and metal tight. The working principle is schematized in Fig. 3.10.

The refractory consumable components include a nozzle housing block situated as an integral part of the ladle lining and designed to match the normal ladle sidewall life. The main working consumable is a metal encased refractory nozzle that is equipped with four or eight lances. This multi-injector tube feature is one of the major design advantages of the system that guarantees that all the required injections, no matter how complex a sequence, can be completed with maximum efficiency, according to the claims of the inventors. It appears to be possible to fit the ISID to virtually any liquid metal containing vessel and for it to become part of any ancillary injection equipment system, either powder or wire feed.

Trials have been carried out at the US Steel Lorain and Gary Works and at BSC Teesside [143]. The claims are reduced reagent costs, improved injectant recoveries, and better reproducibility of treatment as opposed to top lance or wire injections. As is the case with injection through a slide gate, even with ISID, only the future will tell the viability of this approach in day-to-day operation.

### 3.2.1.7 Ladles and Their Refractories

A basic principle, when developing ladle injection, is to design a practical process for the available ladles, with the minimum of alterations. There is, however, always some movement of the steel surface and slag, as well as splashes in the ladle, during injection. Hence, there must be a sufficient freeboard of about 30 cm or more in the ladle. The effect of splashing can also be eliminated by using a ladle cover. This also diminishes heat losses by radiation and makes it possible to remove the smoke and dust generated in the process by means of a suction system.

Ladle injection, like some other ladle treatments, causes very intensive stirring and thus creates conditions in which the rate of reactions between the steel and ladle lining is increased. Therefore, requirements for lining material are much more stringent than for normal ladles.

It is beyond the scope of this presention to discuss ladle refractories in detail. However, the ladle refractory used significantly affects the process, as will be discussed in length later, and therefore this aspect will be considered here with the aid of Ref. 16.

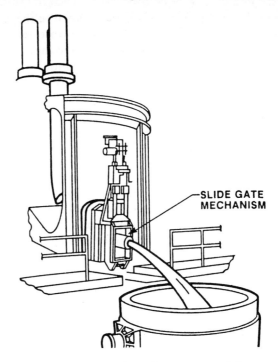

SLIDE GATE
MECHANISM

**Fig. 3.13.** Tap hole slide gate installed on electric arc furnace to control slag carry-over. Courtesy *Iron and Steel Engineer*, January 1986, p. 41.

Tap hole slide gates can be used in open hearth, basic oxygen, and electric arc furnaces. A slide gate mechanism installed on an electric arc furnace is illustrated in Fig. 3.13 [140]. It can be used in two ways: leaving a heel in the furnace, or shutting off the tapping stream at the first sign of slag in the tapping stream. The emergence of slag can be detected either visually or electromagnetically by an electromagnetic level indicator device placed in the tap hole as shown schematically in Fig. 3.14. When metal is flowing through the tap hole, the sensor gives a characteristic signal. Once the material flowing through the tap hole is about 20% slag, the signal changes significantly. One of the problems with such devices is that because of vortexing, slag may be in the center of the metal stream and may cause an ambiguous signal. However, in general, these devices have worked reasonably well in that they can detect slag before visual observation can be made.

As stated, one of the problems associated with slag-free tapping is the vortexing of the steel that develops during tapping and causes a mixture of slag and metal to go through the tap hole [16]. One method of reducing this is excentric bottom tapping of a specially designed electric furnace that only requires a 15–20° tilt, as shown in Fig. 3.15. With this arrangement, there is no furnace launder. The hearth is extended slightly on the tapping side of the furnace, and a tap hole with a mechanical plug is built into this part of the hearth. To tap the furnace, the plug is swung away and the furnace need only be tilted at a small angle to discharge the metal and can be tilted back as soon as slag starts to appear. In addition to

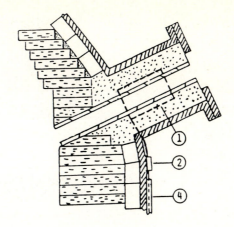

1. SENSOR COILS
2. TERMINAL BOXES
3. HIGH TEMP. CABLING
4. MECHANICAL PROTECTION

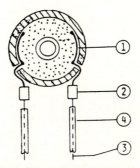

**Fig. 3.14.** Electromagnetic sensor for detecting slag during tap (EMLI slag indicator) [16].

improving slag control, the system is claimed to have other benefits, such as an improved tapping stream giving reduced temperature losses, reduced reoxidation, and nitrogen absorption, in addition to electrical design and refractory benefits.

Still one of the special furnace designs aimed for slag carry-over control is the change of the orientation of the tap hole such that it enters the bath below the slag level more or less tangentially to the arc furnace bottom, see Fig. 3.16 [141]. With liquid heel practice, good results are obtainable with this arrangement.

The Japanese have been [17] experimenting with a teapot-type spout for the tapping of BOF or electric furnaces as shown schematically in Fig. 3.17. Because of the density differences between slag and metal, slag cannot go through the tap hole.

A gas slag stopper is normally used on a basic oxygen furnace, see Fig. 3.18. The furnace is tapped in the conventional manner. At the first sign of slag, an air nozzle is pivoted toward the tap hole and, simultaneously, a blast of high-pressure air is blown into the tap hole, thereby stopping the tapping stream.

Reladling is still another method used to avoid furnace slag. First, a heat is tapped into a ladle in the conventional manner, together with the furnace slag. The steel is then reladled into another ladle through a large diameter nozzle with the reladling

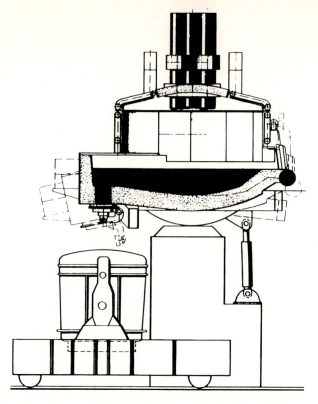

**Fig. 3.15.** Excentric bottom tapping (EBT) electric arc furnace by German Mannesman-Demag.

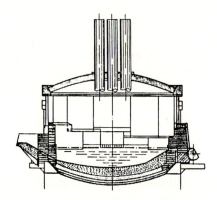

**Fig. 3.16.** Side elevation of an electric arc furnace with low arranged tap hole [141].

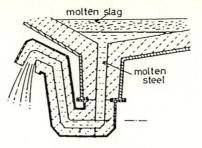

**Fig. 3.17.** Special siphon-type tap hole for slag-free tapping [17].

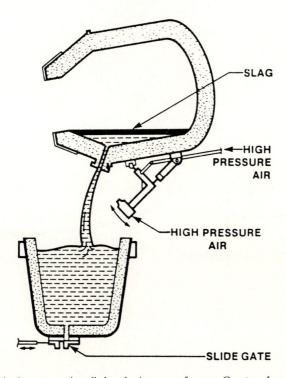

**Fig. 3.18.** Pneumatic slag stopper installed on basic oxygen furnace. Courtesy *Iron and Steel Engineer*, January 1986, p. 42.

operation stopped at the first sign of slag in the reladling stream. This practice, however, tends to cause excessive loss of yield and temperature.

The other alternative of slag elimination is deslagging the ladle after the tap. In its simplest way, this can be done by tilting the ladle and letting the slag flow over the ladle brim. If the slag is sufficiently fluid, reasonable results can be achieved even with this simple practice.

If the ladle is to be deslagged completely in a reproducible manner, the only sure way is to use a slag dragger. Slagging-off the ladle with a dragger calls for:

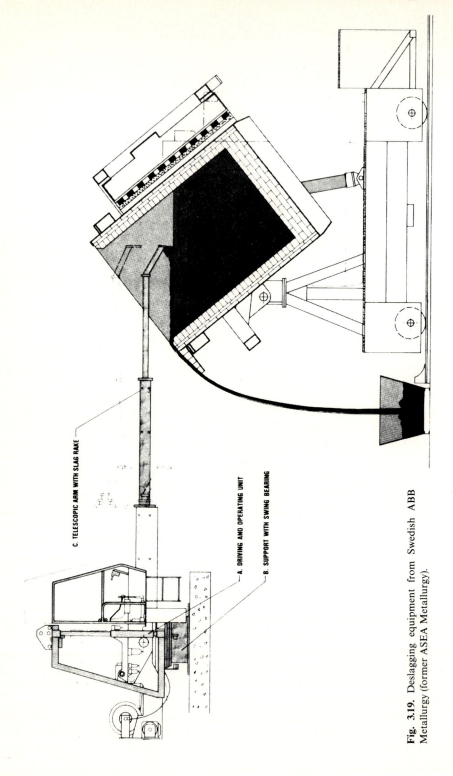

**C. TELESCOPIC ARM WITH SLAG RAKE**

**A. DRIVING AND OPERATING UNIT**

**B. SUPPORT WITH SWING BEARING**

**Fig. 3.19.** Deslagging equipment from Swedish ABB Metallurgy (former ASEA Metallurgy).

1. the possibility of tilting the ladle,
2. a slag rake,
3. a slag spout, and
4. bath movement.

Tilting of the ladle can be done either with the aid of a crane or by using a tiltable cradle.

The slag rake or dragger can be either mechanized or operated manually. Both versions are used. A modern mechanized slag rake is shown in Fig. 3.19. The system consists of a water-cooled rake fastened to a telescopic beam. The beam performs a linear reciprocating movement.

A slag spout may be necessary in order to protect the ladle flange and to prevent the buildup of solidified steel and slag on the ladle brim. A typical slag spout is seen in Fig. 3.19. The spout can either be attached by a hydraulically operated device or placed on and removed from the ladle with the help of a crane.

An appropriate bath circulation is important in moving the slag cover toward the spout. This bath movement can be generated either by gas rinsing through a porous plug or with the aid of an induction stirrer in combination with a ladle furnace. The modern straight induction stirrers have made it possible to tilt the ladle furnace in the ladle car and thereby provide for very simple deslagging. With the stirrer working in the right direction, this is done in a few minutes.

A third possibility for removing the slag from the ladle is to employ suction using a vacuum. With vacuum slag removal, a heat of steel is tapped into a ladle in the conventional manner together with the furnace slag. The ladle is then transported to a deslagging station where a vacuum pipe is lowered above and close to the slag, which is then drawn into the vacuum system.

### 3.2.2.2 Top Slag Employed

The top slag used in the process is very important and must meet certain general requirements. The slag must be highly basic and a sufficient quantity of slag must be present. Generally, up to 10 kg/ton may be required depending on the application. Additionally, the slag has to be fluid for fast reactions and to aid the incorporation of the inclusion particles.

Typically, the top slag will consist of $CaO$, $CaF_2$, possibly $Al_2O_3$, and small amounts of other fluxing agents. The slag should not contain significant quantities of unstable oxides such as $FeO$ or $MnO$. If the amount of these unstable oxides increases, the efficiency of the process decreases, as will be discussed in length later. It follows that the amount of the steelmaking slag that is allowed to enter the ladle has to be minimized.

For best results, the top slag should be added prior to and during tap if the original slag is retained in the furnace. If the ladle is deslagged, the new slag has to be charged after the skimming has been completed.

### 3.2.2.3 Typical Operation of the Injection Equipment

To start an injection cycle in typical current practice, powder of predetermined weight is transferred from the storage container to the dispenser by opening the connecting valves. When the predetermined weight is registered on the weighing system, the valves are closed. The weighing system is then set to the desired amount of powder to be injected in the steel. The system is then ready for injection.

The ladle, full of steel, is placed on the injection pad. The hood, if employed, is set on top of the ladle with, for example, an overhead crane, making sure that the injection and exhaust holes are properly aligned. The lance stand arm is then raised and swung to bring the lance over the injection hole in the hood. The lance is lowered to a distance of about 15 cm from the hood. The transport of powder and carrier gas is initiated and, as soon as a smooth, uniform flow is achieved, the lance is lowered into the bath and the exhaust damper is opened. The injection is supervised by an operator who can manually adjust the pressures and the flow rates if required.

When the predetermined weight of powder has been injected, the powder flow is automatically cut off, but the gas flow continues until the lance is withdrawn from the steel. The lance is then raised and swung over to the lance magazine, and the exhaust damper closed automatically. The hood is then removed and the ladle taken to the casting bay.

In examining in detail what happens in the dispenser during injection, one must first create a driving force to propel the powder out of the dispenser into the transport gas stream, that is, the pressure differential has to be established by increasing the pressure at the top of the dispenser. This forces the fluidized powder through the outlet orifice. The flow rate of the powder, kg/min, is controlled mainly by the pressure difference between the top of the dispenser and the ejector, the outlet orifice diameter, and the size of the lance nozzle.

An adequate value for the pressure difference quoted previously is $\geqslant 2.0$ bar, which should guarantee a successfull injection down to powder flow rates of, for example, 10 kg/min, even in case of slight variations or disturbancies in the powder feeding rate.

The characteristic features of an injection system affecting powder injection are to some extent interchangeable and can be presented schematically in a graph. An example of such a graph for operational parameters of one particular injection system is given in Fig. 3.20 [18]. As can be understood from this figure, the desired powder feeding rates can be achieved with numerous different combinations of the operating parameters of the system.

As regards the flow rate of the carrier gas, an unnecessary high flow rate should be avoided. If the gas/powder ratio is sufficiently low, smaller gas bubbles will form in the melt. This, with the average particle size of the powder, will increase the amount of powder that penetrates through the gas bubbles into the melt. Consequently, the efficiency of the injection will increase with the increasing surface area of powder in contact with the melt.

Furthermore, the feeding rate of the powder should not be too high as the steel cannot adsorb very large amounts in a given time. The values in Table 3.2 give guidelines for optimum calcium injection rates depending on the bath temperature and the type of calcium carrier.

For example, considering a 60-ton ladle, the maximum feeding rate of CaSi would be roughly 20 kg/min. If the feeding rate is greater than this value, the calcium may be partially vaporized and will subsequently burn on the surface of the melt.

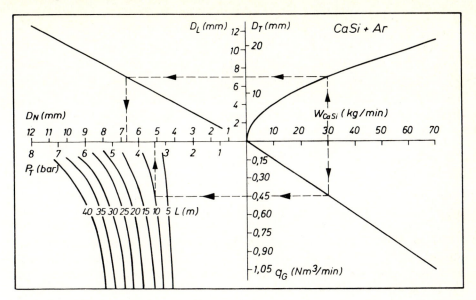

**Fig. 3.20.** Operational parameters of injection system in Slovenian Steelworks-Store [18]: $D_L$, $D_N$, and $D_T$ are diameters of lance nozzle, dispenser outlet orifice, and transport pipe, respectively; $P_T$, ejector pressure; $L$ transport length; $q_G$, quantity of transport gas; $W_{CaSi}$, powder feeding rate.

**Table 3.2.** Guidelines for optimum calcium injection rates[a]

|  | Range of temperature (°C) | | |
| --- | --- | --- | --- |
| Products | 1580/1600 | 1600/1620 | 1620/1640 |
| Pure calcium granules (blended) | 60 ppm of calcium per min | 50 to 60 ppm of calcium per min | 40 to 50 ppm of calcium per min |
| Calcium aluminium granules (blended) | 90/100 ppm of contained calcium per min | 80/90 ″ ″ ″ | 70/80 ″ ″ ″ |
| Calcium nickel granules (blended) | 90/100 ppm of contained calcium per min | 80/90 ″ ″ ″ | 70/80 ″ ″ ″ |
| Calcium silicide comparison | 120/130 ppm of contained calcium per min | 100/120 ″ ″ ″ | 100 max ″ ″ ″ |

[a] From Jehan and Aquirre [135].

```
Factors affecting the injection of powdered materials in steel

  1  Type of material being injected (metals, carbides,
     silicides, oxides etc.)
  2  Grain size and structure of reagent
  3  Type, quantity and loading of conveyor gas
  4  Powdered materials injected and injection time
  5  Injection depth
  6  Oxygen content of steel
  7  Ladle size and form
  8  Type of ladle refractory material
  9  Slag composition and quantity
 10  Oxygen pick-up from the atmosphere
```

**Fig. 3.21.** Factors involved in the injection treatment of steel [7].

For proper results, the positioning of the injection lance is important. One has to understand that the deeper the immersion of the lance, the better will be the yield of the injected alloys. The same applies to the mixing time, that is, the time required to homogenize the melt with respect to the analysis of the added element. The mixing time with 10% lance immersion is roughly three times longer than for a 95% immersion.

In case of CaSi injection, the minimum lance immersion to prevent calcium from vaporizing immediately can be described by the following simple formula:

$$h = 1.47(10^a - 1)$$

$$a = 4.55 - \frac{8027}{T}, \tag{3.1}$$

where $h$ gives the immersion in meters and $T$ is the bath temperature in degrees Kelvin.

Regarding the lance position in the horizontal plane, there are at least two factors to be considered. If the lance is positioned next to the ladle wall, this will give the shortest possible mixing time of the bulk. However, this normally results in skulling of the ladle mouth. The central position of the lance yields a longer mixing time as opposed to the previous case, but may help in reducing the skulling of the ladle, possibly even in decreasing the refractory wear. The suitable practice remains to be established for each individual case. Factors that are generally considered to affect injection results are summarized in Fig. 3.21 [7]. Consequently, at least these have to be taken into account when considering injection treatment.

The following typical example, updated from the metallurgical literature [16] will summarize the events during a injection treatment for desulphurization. The process is for a steel containing 0.02% to 0.03% sulphur with an aim of a final sulphur of less than 0.005%.

1. Some type of slag retention should be used in avoiding excessive steelmaking slag carry-over. Details of these methods are discussed in Section 3.2.2.1.
2. The steel must be thoroughly deoxidized. Oxygen sensor measurements in the furnace may be helpful in controlling deoxidation.

3. The synthetic top slag should be added prior to or during tap. For example, CaO and $CaF_2$ in the ratio 80 : 20 or 85 : 15 can be used.
4. Depending on the success of the slag retention, it may be necessary to measure the amount of steelmaking slag carry-over at this point. If the amount is excessive, for example, greater than 10 cm, additional Al should be added to reduce the slag.
5. An oxygen sensor measurement may be helpful at this point to predict the amount of desulphurization agent, for example CaSi, required. This will depend on the desired degree of desulphurization and possibly on the oxygen activity in the metal prior to injection. Typically, about 2.5 kg of CaSi per ton is injected.
6. After injection, a gentle argon stirring for a couple of minutes will help float out inclusions.
7. Trim additions of alloys may be needed after the injection.

There are, of course, many variations on this typical process because of special considerations. For example: (1) a flux may be injected in place of CaSi, (2) CaO may be injected before CaSi to produce a stiff top slag if a ladle cover is not used, and (3) smaller amounts of CaSi can be used if only oxide modification is desired for casting purposes instead of desulphurization.

### 3.2.2.4 Injected Materials

Normally the aim of the metallurgical treatment prescribes the main chemistry of the agent to be injected. Further constraints may be put forward by economy, availability, safety, handling aspects, yield, and the like.

Regarding injection, the following powder characteristics are thought to be generally desired:

1. mean grain size, $\sim 0.1$ mm;
2. maximum grain size, $\sim 1.0$ mm;
3. loss on ignition, $< 1.0\%$; and
4. shape, round or cubic.

Fine powders are more reactive but excessive fines should be avoided because of their tendency to cause difficulties in pneumatic conveying. Examples of the principal powders injected in practical steelmaking to accomplish different objectives are given in Table 3.3.

The most common powders currently injected into steel are CaSi and lime-based products, normally $CaO + CaF_2$. The normal quality CaSi contains 30% calcium. This CaSi is used for desulphurization, deoxidation, and to modify oxide and sulphide inclusions.

The $CaO + CaF_2$ mixture can be used as an alternative to CaSi for desulphurization and deoxidation (lowering of the total oxygen content). The proportions of CaO to $CaF_2$ in the mixture are normally 80 to 20 or 85 to 15.

It is generally known by every steelmaker who has tried to inject lime-based materials that this may pose serious practical difficulties. Therefore, a short discussion about injectable lime may serve a purpose [19]. The requirements for lime-based material to be injectable can be summarized as follows:

**Table 3.3.** Examples of powders applied for injection

| | |
|---|---|
| Dephosphorization: | $CaO + CaF_2 + Fe_2O_3$<br>Soda ash |
| Desulphurization: | $CaO\ (+CaCO_3)$<br>$CaO + Al$<br>$CaO + CaF_2\ (+Al)$<br>Soda ash<br>$CaC_2\ (+CaCO_3, CaO)$<br>$Mg\ (+CaO, Al_2O_3)$<br>Misch metal |
| Deoxidation: | Al<br>CaSi, CaSiBa ⎱ Modification of<br>CaSiMnAl ⎰ inclusions |
| Alloying: | $FeSi, CaCN_2, C$<br>$NiO, MoO_2$<br>FeB, FeTi, FeZr, FeW, SiZr<br>Pb, FeSe, Te |

1. flowability,
2. low moisture content and low hygroscopicity, and
3. high basicity and sulphur capacity.

There are also requirements that vary according to particular steel plant conditions, such as melting rate, fluidity, and deoxidizing capacity. Other considerations are the environmental ones, such as the amount of dust and fume generated by the product. Flowability is, of course, of prime importance because without this property it is impossible to inject. Size grading and particle shape are important from this point of view but what must also be taken into account is the friction between the particles. In this respect, the surface of the soft burned lime (calcined at 1000°C) is much more porous than that of the hard burned lime (calcined at 1400°C).

These surface properties are also important from the point of view of reactivity and susceptibility to moisture absorption. In storage, following calcining, lime will naturally react with moisture and carbon dioxide. These reactions and their products reduce the efficiency of desulphurization and result in increased hydrogen contents of the steel. A higher porosity gives a greater effective surface area available for reaction with atmospheric moisture and has been found to have a far greater effect than particle size on the moisture absorption of a particular lime batch during the first days of exposure to the environment.

An allied effect of surface properties is on the flowability. The reaction products increase the tendency of the lime to snowball, that is, the particles will tend to stick together, become more difficult to fluidize, and have an increased tendency toward blockage.

The increasing importance of injection and the very poor and inconsistent injection properties of the vast majority of limes, together with increasingly stringent hydrogen specifications, have prompted various suppliers to investigate the fundamental properties of lime and lime-based products and to develop techniques to produce products with minimal moisture content and hygroscopicity, together with excellent injectability.

### 3.2.2.5 Safety Aspects

Working with powders of reactive materials requires special precautions because of risks of dust explosions. A dust explosion is the fast self-sustaining reaction when small dust particles are suspended in air and rapidly burn under heat evolution. Practically all combustible materials and some materials that are normally regarded as noncombustible might under certain circumstances cause explosions if they are sufficiently fine in size and mixed with air in a certain proportion range.

Therefore, precautions have to be taken when handling the injection equipment and the powdered materials, particularly during maintenance work. If a small quantity of dust is entrained and ignited, that primary explosion may in itself be very limited. It might, however, cause a secondary and much more severe explosion by ignition of much larger quantities of powdered material whirled up by the primary explosion.

The tendency of powered material to explode is connected with the following factors:

1. particle size,
2. particle structure,
3. the quantity of particles in the air,
4. its reactivity to oxygen,
5. content of noncombustible material, and
6. required energy to ignite a dust–air mixture of the material.

Dust, in this case, consists of particles of inorganic or organic material that can be suspended in the air and under certain circumstances may give a dust explosion. Generally, particle size below 0.2 mm (for some materials < 0.5 mm) is regarded to entail explosion risks. Particularly the finest dust, < 0.03 mm of certain materials, is to be considered as very dangerous, except when they are diluted with coarse material.

Dust particles of angular shape in teeth form, wirelike, or in flakes—in other words with sharp corners—have been shown to have much higher explosion risks than round or oval particles. A smooth, hard surface also results in a reduced explosion.

In defining the risks, these definitions are used: minimum and maximum explosion limit concentrations and the explosion index, respectively.

The minimum concentration of dust particles in the air when the material can be brought to an explosion is the minimum explosion limit. The maximum explosion limit is the maximum concentration of dust particles in the air when the material cannot explode. The explosion limits are normally given in $g/m^3$.

The explosion index $E$ is a measure of the risks for dust explosions or rather how severe the explosion is when it occurs. The index $E$ has been defined as follows:

$E = 10$           Severe risk for explosion
$10 \leq E < 1$    Strong risk for explosion
$1 \leq E < 0.1$   Moderate risk for explosion
$0.1 \leq E < 0$   Weak risk for explosion
$E = 0$            No risk for explosion

The explosion indices of various materials have been summarized in Table 3.4 [20]. The indication from Table 3.4 is that special precautions are needed when conveying fine Mg and Al powders. One should note, however, that the powders used in the injection technique have greater particle sizes and are therefore less dangerous.

To summarize, the safety rules for the manipulation of the most common powder in injection metallurgy, CaSi, are given here. All premises on which CaSi powder is used or stored should be rigorously clean and steps should be taken to avoid any accumalation of CaSi dust. In all cases, if a dust cloud is formed, systematic steps must be taken to ensure the removal of

1. any electrical, static, dynamic, or mechanical source of sparks (contactors, electric arcs, grinding, sealing pistols, etc.); and
2. incandescent filaments, drops of molten metal, or naked flame.

The gases used in any pneumatic transport system must be inert, or at least have an oxygen content of less than 9%. It is recommended that empty containers should be stored in the open air before reuse or remelting. Never carry out welding operations in the presence of a dust cloud or on a receptacle containing power. Never scratch these containers with metal objects (use wood).

CaSi powder is very difficult to burn when stocked in piles. If it does burn, the combustion is slow and spreads by degrees. Under no circumstances should fluids under pressure, such as fire hoses and emulsion extinguishers, be used to extinguish

**Table 3.4.** Potential for explosion of selected powders[a]

| Powder | Particle size | Minimum explosive concentration (g/m$^3$) | Minimum ignition temperature clouds (°C) | Minimum ignition energy (J) | Explosion index $E$ |
|---|---|---|---|---|---|
| Al flakes | 100%–44 $\mu$m | 45 | 650 | 0.020 | >10 |
| Al atomized | 77%–44 $\mu$m | 45 | 740 | 0.080 | 0.9 |
| Al atomized | 6 $\mu$m | 50 | 820 | 0.050 | >10 |
| Mg ground | 77%–74 $\mu$m | 40 | 620 | 0.320 | 0.3 |
| Mg ground | 100%–100 $\mu$m | 25 | 620 | 0.040 | >10 |
| CaSi | 2 $\mu$m | 60 | 540 | 0.150 | 2.0 |
| CaSi | 100%–74 $\mu$m | 90 | 690 | 0.220 | 0.2 |
| CaSi | 78%–74 $\mu$m | 285 | 870 | | |
| FeSi 25/75 | 100%–74 $\mu$m | 400 | | 0.400 | |
| FeMn 1% C | 100%–74 $\mu$m | 130 | 450 | 0.080 | 0.4 |
| FeCr 69% Cr 5% C | 95%–44 $\mu$m | 2000 | 790 | | |

[a] From Wikander [20].

any fire, since such action may create a dust cloud. The recommended procedure in case of fire is to smother the combustion with dry and inert powders, such as sand, dolomite, magnesium oxide, calcium oxide, or calcium fluoride.

## 3.3 Theoretical Approach to Powder Injection

### 3.3.1 Possible Zones in the Ladle

The situation in the ladle during powder injection is sketched in Fig. 3.22 [21]. The ladle can be divided into the following hypothetical zones:

1. jet zone, in front of the lance outlet;
2. major pump zone (bubble zone), formed by the rising gas bubbles and injected material;
3. breakthrough zone, where gas bubbles emerge through the slag into the atmosphere;
4. slag zone;
5. dispersion zone, where the injected gas and slag and even top slag can be dispersed;

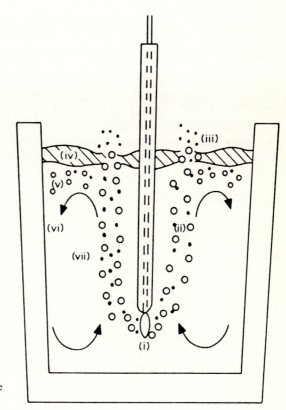

**Fig. 3.22.** Possible zones in the ladle during powder injection [21].

6. lining zone, where metal is in contact with the lining; and
7. intermediate zone, with lowest stirring intensity.

A short discussion of some selected zones is presented in the following sections.

### 3.3.1.1 Jet Zone

Downstream of the lance nozzle, the gas powder mixture penetrates into the melt. The penetration depth is determined by a balance between the inertia and the buoyancy forces. The following values can be considered typical for the vertical penetration of the injected powder [22]:

| | | |
|---|---|---|
| material flow rate | 20 | 40 kg/min |
| gas flow rate | 500 | 500 dm$^3$/min |
| ferrostatic pressure | 1 | 1 bar |
| vertical penetration | 25 | 19 cm |

Considering further the area downstream of the lance nozzle, the injected particles can be brought into contact with the melt by different mechanisms:

1. penetration into the melt at the gas–liquid boundaries because of inertial forces,
2. collection on the metal drops in the froth region of the jet (venturi scrubber principle),
3. deposition at the gas–liquid interface because of centrifugal forces when the main gas flow direction is inverted (cyclone principle),
4. deposition at the rear interface of the bubble by gravitational settling, and
5. deposition at the bubble–liquid interface because of internal circulation.

In spite of considerable research, very little is known about the different regimes and mechanisms. The only thing that is sure is that the injection of powders is a very effective method of contacting metallurgical reactants.

For case 1 (inertial penetration of particles through gas–liquid interfaces), the two balancing forces at the interface are surface tension and inertial forces. Following a simple model, the critical particle size for penetration of the bubble–melt interface has been calculated for carbon particles [23]:

$$r_\mathrm{p} = \frac{2.6\,\sigma_\mathrm{gm}}{U_0^2},\tag{3.2}$$

where $r_\mathrm{p}$ is the critical particle radius, $\sigma_\mathrm{gm}$ is the interfacial tension at the gas–melt interface, and $U_0$ is the initial velocity of the particle relative to the interface.

The critical particle radius, that is, the size of the particles that can penetrate into the melt, is shown as a function of the velocity in Fig. 3.23 [22]. According to this model, most of the particles would have enough kinetic energy to penetrate the interface, but the situation is not quite as simple both because of turbulence in the gas phase and drops and bubbles.

There have been numerous papers on this topic [23]. It has been concluded [24] on the basis of examinations in water models, using alumina or styrene particles to simulate the injected particles, that Mg particles of 2 mm in diameter injected with nitrogen gas at 3 Nm$^3$/min could penetrate into the melt. The validity of Eq. (3.2)

**Fig. 3.23.** Penetration limit for carbon particles [22].

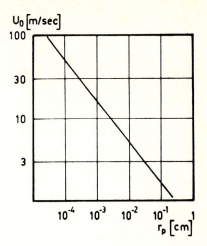

**Fig. 3.24.** A model representing particle distribution in an argon bubble in steel [26].

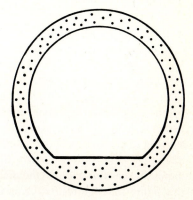

was also confirmed by similar water model experiments [25] calculating through the equation that 65% of injected Ca-alloy particles penetrated into the melt and the rest adhered to the gas–metal interface, as shown in Fig. 3.24 [26]. After hot model experiments, an interesting model was proposed [25] in which Ca-alloy particles brought into the bulk melt and those adhered to the gas–metal interface remain, respectively, liquid and solid while floating up. This is quite in contrast to an earlier model [27] considering that Ca-alloy particles injected in the melt evaporated into bubbles.

### 3.3.1.2 Major Pump Zone

The expanding gas and the rising bubbles in the major pump zone induce a circulation in the bath that is described as a "mammoth pump." The mammoth pump is frequently used, for example, in chemical engineering, for ice prevention in harbors, and in metallurgy.

Generally, it can be said that this zone is the engine of the whole circulation brought about by the injected gas. To give an example, in an actual case, argon gas at a flow rate of 600 dm$^2$/min was introduced into the bath at a lance immersion depth of 90 cm. By assuming a thermal expansion factor of 6.3, the mammoth pump capacity was calculated [22] to be 10.5 tons/min. At the same time, the circulation rate due to thermal natural convection, that is, the circulation in absence of any gas, was measured [22] by tracer addition in a 6-ton ladle to be of the order of magnitude of 1.5 tons/min.

The kinetic energy contained in the flowing liquid leaving the mammoth pump at the surface of the melt is dissipated in the bulk of the bath (turbulent decay). Thus, mixing occurs by forced convection and turbulence. The following empirical correlation between mixing time and dissipated energy in metallurgical reactors has been reported in the literature [28] and is discussed in Chapter 2.

$$\tau_b = 800 \, \dot{\varepsilon}^{-0.4}, \tag{3.3}$$

where

$$\dot{\varepsilon} = 0.014 \frac{VT}{G} \ln\left(1 + \frac{h}{1.48}\right),$$

$\tau_b$ is the mixing time (s), $\dot{\varepsilon}$ is the dissipated energy (W/ton), $V$ is the gas flow rate (dm$^3$/min), $T$ is the temperature of the melt ($K$), $G$ is the mass of the melt (ton), and $h$ is the melt height above the gas outlet (m).

The schematic presentation of this correlation can be found in Fig. 3.25 [28]. Only part of the power input is dissipated in the bath and used there to mix and to promote the coalescence of inclusions. The level of utilization of available power decreases with increasing specific gas flow rates and further depends on the immersion depth and design of the lance. The decrease of the available power as a function of the gas flow rate is shown in Fig. 3.26, which demonstrates the effect of the increasing gas flow rate on the mixing time of a melt. It can be readily deduced from this figure that the returns from an increase in the gas flow rate are diminishing when increasing the specific gas flow rate. It is to be noted further that if powder is

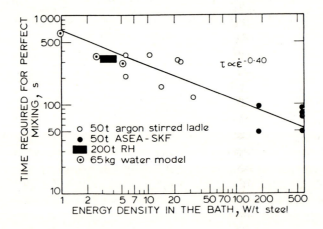

**Fig. 3.25.** Relationship between the time required for perfect mixing ($\tau_b$) and the dissipation of energy in various steel processing operations [28].

injected with the gas the mixing time is not affected to any greater degree as opposed to gas injection only [134].

It can be stated that the best conditions for reactions are in the bubble column of the major pump zone at the metal–gas phase boundary and on the ladle wall. This will be seen when the deoxidation and desulphurization processes are discussed.

A basic principle of ladle injection is to force the reactive material deep into the steel melt. The injected material, either in solid, liquid, or gaseous form, reacts with the surrounding steel as it rises through the steel melt in the bubble column. This basic idea was the starting point of calcium- and slag-injection treatment research.

The process can be treated with the aid of the reactor models presented already in early 1960s by Schenk [29, 30]. Accordingly, there is the so-called transitory phase contact between the injected material and the steel melt in the bubble column during ladle injection, see Fig. 3.27 [29, 30]. When the model is used for desulphurization, the following formula is obtained:

$$\frac{[S]}{[S]_0} = \exp(-\eta\gamma), \tag{3.4}$$

where $[S]_0$ is the initial sulphur content, $[S]$ is the sulphur content after the treatment, $\eta$ is the partition coefficient of sulphur (S)/[S] after the treatment, and $\gamma$ is $m_s/m_m$, the mass ratio (slag/metal).

For a specific injection time, the following equation is valid:

$$\frac{[S]}{[S]_0} = \exp\left(-\eta\frac{\dot{m}_s}{m_m}t\right), \tag{3.5}$$

where $\dot{m}_s$ is the slag injection rate and $t$ is the time.

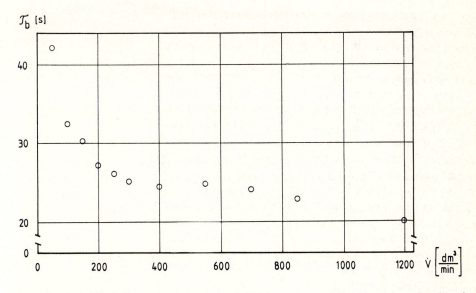

**Fig. 3.26.** An example of the decrease of the mixing time (in a water model in this case) as a function of the gas flow rate. Reprinted with permission from [35].

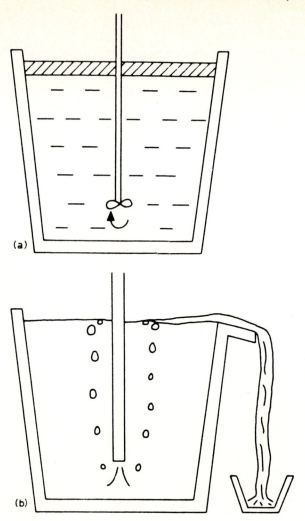

**Fig. 3.27.** Principle of (a) a two-phase permanent reactor and (b) a transitory reactor. Redrawn from [29, 30] by [21].

Transitory phase contact is very different from permanent phase contact, which prevails between the steel and top slag during gas bubbling. When desulphurization proceeds (in the latter case), the sulphur content of the top slag increases all the time, and finally approaches the equilibrium distribution. The extent of desulphurization is then determined by

$$\frac{[S]_0}{[S]} = 1 + \eta^* \gamma. \tag{3.6}$$

Comparison of Eqs. (3.4) and (3.6) shows that, with the same amount of slag, transitory phase contact gives better desulphurization than permanent phase con-

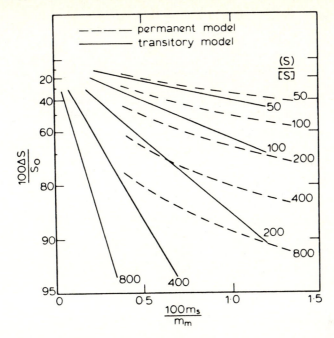

**Fig. 3.28.** Dependence of the degree of desulphurization $\Delta S/S_0$ on the theoretical relative slag amount $m_s/m_m$ with different sulphur distributions (S)/[S] [146].

tact (Fig. 3.28 [146]). In the permanent reactor model, the sulphur content of the slag increases until the equilibrium sulphur distribution, or in some cases sulphur saturation, is attained.

In the transitory model, the slag droplets that form can reach sulphur saturation or the maximum distribution value during their passage through the steel melt. It has been claimed [31] that these slag droplets rising in the steel bath can contain up to 8%, possibly even more, sulphur in the steel having oxygen activity of 4 ppm and sulphur content above 0.020%. With injection, new material with high sulphur capacity is continuously brought into the melt, therefore desulphurization can proceed much further. This mechanism can work only if the floating slag is removed from the system after its reaction. In practice, this does not happen, and the injected slag remains in the top slag. Additionally, it has been stated [31] that the transitory reactor only starts to work when the sulphur content has come down to low levels. Hence, ladle injection is a type of mixed reactor, as shown in Fig. 3.29 [32].

### 3.3.1.3 Slag Zone

The top slag has several important functions, for example, (1) absorption of the reaction products, (2) chemical isolation, and (3) thermal isolation.

Regarding point (1), inclusion deposition on the top slag, for both turbulent transport and flotation, the inclusion diameter is an important factor. In general, to reach the slag, the inclusions dispersed in the moving melt must have enough

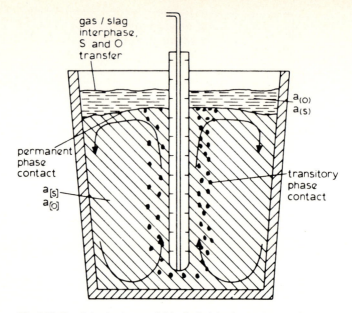

**Fig. 3.29.** Desulphurization model for ladle injection under top slag [32].

momentum to penetrate the laminar boundary layer between the melt and the slag. Such a laminar boundary layer near liquid surfaces has a thickness of roughly 4 mm [33].

The deposition to the slag through the laminar boundary layer can also take place by an other mechanism: gravitational rise. The particles rise with a velocity of $10^{-4}$ (2 $\mu$m) to $10^{-2}$ (20 $\mu$m) cm/s. The contacting time of some 2–3 s is enough for bigger particles, $d_p \geq 20$ $\mu$m, to reach the slag.

Another important possible reaction in the slag zone is redispersion of slag. Still today, very little is known about this phenomen in liquid steel–slag contacting. High turbulence intensity in this boundary zone will result in redispersion. At least the following parameters will influence the redispersion rate:

1. slag composition and temperature (fluidity, density, and surface tension), and
2. bath movements (turbulence intensity, wave movements, and wave breaking).

In addition, the slag plays an important role in preventing the access of oxygen to the melt (2) and will also act as a thermal insulator assisting in decreasing the temperature losses of the steel (3).

### 3.3.2 Rate-Determining Steps in Injection

There are three major rate-determining processes: powder feeding, reaction at the interface, and mixing of the bulk phase during powder injection [34], as is schematically shown in Fig. 3.30, in which these rate-determining processes are represented by the following three factors, all of which have dimensions of time:

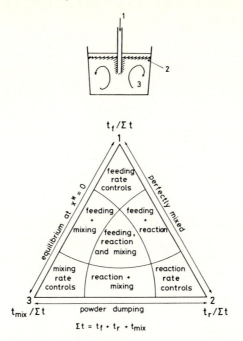

**Fig. 3.30.** Schematic diagram showing rate-determining processes during powder-injection refining. An explanation of the vertices is given in Table 3.5 [34].

1. The powder-feeding rate is represented by the time, $t_f$, necessary to feed a sufficient amount of powder. It is given by

$$t_f = \frac{W_m}{P_{in} L_{wt}}, \tag{3.7}$$

where $W_m$ is the mass of the metal phase, $P_{in}$ is the powder injection rate, and $L_{wt}$ is the sulphur partition coefficient.

2. For the reaction at the interface, the reciprocal of the rate constant $K$ is chosen, and expressed as the time, $t_r$, required for sufficient reaction, that is,

$$t_r = K^{-1} = \frac{V_m}{A_r k}, \tag{3.8}$$

where $V_m$ is the volume of the metal phase, $A_r$ is the area of the reaction at the interface, and $k$ is the mass-transfer coefficient.

3. For the mixing process, the bulk mixing time, $t_{mix}$, is chosen.

These factors, which are summarized in Table 3.5, are useful in a qualitative consideration of the kinetics during powder injection.

The ternary diagram in Fig 3.30 shows a schematic representation of the concept of the rate-determining processes being controlled by these three factors. Vertex 1 ($t_f \gg t_r + t_{mix}$) of the ternary diagram corresponds to conditions in which powder feeding alone determines the total reaction rate, while vertices 2 and 3 correspond

**Table 3.5.** Rate-determining processes during powder injection[a,b]

| Vertex | Rate-determining process | Factor |
|--------|--------------------------|--------|
| 1 | Powder feeding | Time to feed sufficient amount of powder: $t_f = W_m / P_{in} L_{wt}$ |
| 2 | Reaction or mass transfer in vicinity of reaction interface | Time for reaction: $t_r = V/Ak$ |
| 3 | Mixing of bulk phase | Time for mixing: $t_{mix}$ |

[a] From Ohguchi and Robertson [34].
[b] See Fig. 3.30.

to circumstances in which reactions at the interface or bulk-phase mixing, respectively, solely determine the rate. For vertex 1, it has been shown [34] that for $t_f/t_r > 1.2$, powder feeding controls the kinetics if $t_{mix,m}$ is sufficiently small.

## 3.4 Discussion of the Results Obtained with Powder Injection

### 3.4.1 Desulphurization

It is obvious that different injection techonological parameters have a noticeable influence on the metallurgical results of injection treatment.

*The lance immersion depth* has been shown [2] to influence desulphurization results in calcium carbide injection. The achieved desulphurization with full immersion of 270 cm was 55% and with a 50cm immersion fell to 30%. These results were confirmed later [144] by the injection of both CaSi and CaO + CaF$_2$ with full immersion and shallow immersion of the lance in absence of simultaneous gas stirring, see Fig. 3.31. As far as the shallow injection was concerned, the degree of desulphurization was clearly drastically lower as opposed to the case of full immersion of the lance. Diffusion-controlled material transport has been suggested [2] as an explanation for the above phenomenon. On the other hand, the general stirring effect in the ladle has been shown [13, 35] to be essentially weaker with shallow injection, as opposed to deep injection. As was shown in Section 3.3, the mixing rate is one of the controlling factors of for example, desulphurization.

Different types of lance nozzles have also been tested [13, 36]. Multihole nozzles gave more finely dispersed gas bubbles and resulted in a somewhat better stirring efficiency than a one-hole nozzle, with the result that better desulphurization efficiency could be expected from the injection treatment.

*The feeding rate* of the injected powder affects the extent of desulphurization, as has been already discussed. It has been shown [37] that a reduction of the specific feeding rate of CaSi powder from around 20 g/ton · s down to, for example, 5 g/ton · s will double the desulphurization degree under otherwise similar conditions. This matter has also been discussed in Section 3.2.2.3. However, the situation is different when injecting slag powder, such as CaO + CaF$_2$. There is a lot of evidence [13, 34, 38] showing that a higher feeding rate in this case will improve the desulphurization, see Fig. 3.32 [38]. The probable explanation for the improvement in desul-

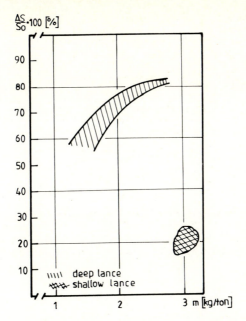

**Fig. 3.31.** Comparison of obtainable desulphurization degrees with deep and shallow lance immersions, respectively [144].

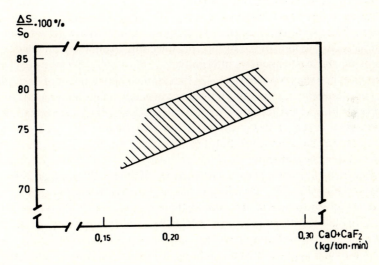

**Fig. 3.32.** Desulphurization degree after injection of 1.75 kg CaO + CaF$_2$, at different rates, per ton of steel [38].

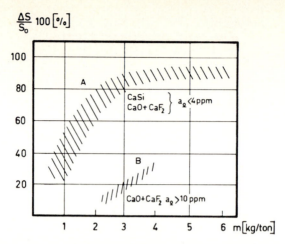

Fig. 3.33. Desulphurization as a function of the amount of powder injected as well as oxygen activity of the steel [40].

phurization performance is that it is produced by the increase in the relative contribution of the transitory reaction in the desulphurization process.

*Injected material*, CaSi, $CaC_2$, or $CaO + CaF_2$, has no or very little influence on the obtainable extent of desulphurization. There is sufficient proof [1, 37, 38, 39, 40, 44, 46] to claim that equal degrees of desulphurization can be obtained with at least all of these powders as a function of the amount of powder injected, see Fig. 3.33. The prerequisite for the equal degree of desulphurization is a low activity of oxygen, for example, below 4 ppm, in the steel before injection. From these powders, $CaO + CaF_2$ is by far the least expensive alternative.

However, it has been reported [41, 42] that a much better extent of desulphurization can be obtained with the same amount of injected magnesium metal compared with CaSi, see Fig. 3.34. Alternatively, for the given extent of desulphurization, much less magnesium is required. Generally, it can be said that the necessary reagent quantities cannot be generalized but depend greatly on prevailing conditions and are specific for each melt shop.

*Slag powder* for injection is usually a mixture of lime and fluorspar, the amount of fluorspar being 10–25%. Desulphurization improves when pure lime is fluxed with $CaF_2$ [43]. With these compositions, there is at least 45% of CaO undissolved at 1600°C [37]. Accordingly, the injected slag powder does not form molten drops, but there are undissolved lime particles, possibly surrounded by liquid slag. Obviously, with such slightly fluxed mechanical powder mixtures, reactions in the injection plume will proceed at a rather low rate.

On the other hand, experience has been gained with the injection of premelted slag powders [31, 36, 44, 45]. Slags of 48% CaO–52% $Al_2O_3$ and 65% CaO–23% $Al_2O_3$–12% $SiO_2$ composition, which are liquid at 1600°C, have been tested [31, 32]. Desulphurization reaction between the steel melt and injected slag droplets along the transitory phase contact path could thus be established. However, no

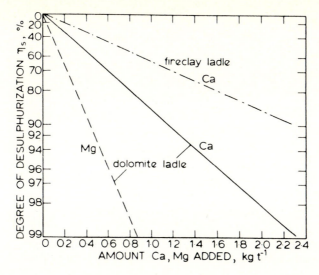

**Fig. 3.34.** Amount of Mg, with respect to Ca, for desulphurization in 120-ton ladles [41]. Reprinted with permission from E. Spetzler and J. Wendorff, Radex Rundschau 1, 595–608 (1976).

distinct difference in injection results between the mechanical mixtures and pre-melted $CaO–CaF_2$ or $CaO–CaF_2–Al_2O_3$ slags has been observed [44]. Evidently, the role of top slag was dominant in this case, and the reduction of top slag and the growth of its sulphur capacity were actually the rate-determining factors.

*Stainless steel* can also be desulphurized with injection techniques. An average desulphurization extent of 54% has been reported [47] with injection of 2.5 kg of 80% CaO + 10% $CaF_2$ + 10% Mg mixture per ton of steel. The starting sulphur level was 0.025%.

*The ladle refractory* has been observed to have a great influence on desulphurization results [2, 7, 37, 48, 49]. The extent of desulphurization is higher when $SiO_2$-containing lining material has been replaced by high alumina or dolomitic lining [7, 37], as already indicated in Fig. 3.34. A silica-containing acid lining is not stable; it decomposes and causes the oxygen activity to increase in the steel melt. This has been established by oxygen activity measurements represented in Fig. 3.35 [50]. The oxygen activity is high on the ladle wall because of the decomposing reaction of silica. However, even at the center of the ladle, the measured values are remarkably higher than the Al–O equilibrium, which could be established in the dolomitic ladle. The distribution of oxygen activity is illustrated in Fig. 3.36(a) for a gas bubbling process in an acid ladle. This must be considered as the presentation of the progress of reoxidation rather than as a real map of oxygen activity gradients in the ladle. During injection, the situation is quite analogous on the ladle wall, see Fig. 3.36(b). Although the oxygen activity is lowered in the bubble zone, in this case by the injected material, as opposed to Fig. 3.36(a), very low sulphur levels cannot be attained. Generally, it can be stated that in acid ladles a sulphur level of 0.008% or less is achievable [2, 37, 48], but for very low sulphur, for example, below 0.004%, a basic lining is mandatory. A comparison of desulphurization results in acid and

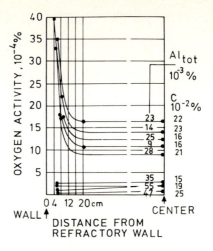

**Fig. 3.35.** Oxygen activity distribution in acid ladle (upper curves) and dolomitic ladle (lower curves) [50].

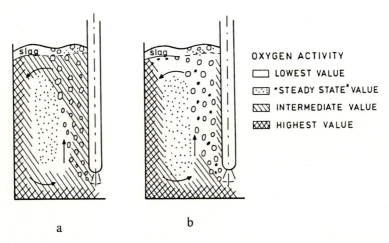

**Fig. 3.36.** Schematic model of oxygen activity distribution in an acid ladle: (a) gas bubbled ladle and (b) powder injection [51].

basic ladles is shown in Fig. 3.37 [2]. The difference in the desulphurization extent obtained because of the refractory material of the ladle is clear.

*Unstable oxides* in the top slag, such as FeO and MnO, will decrease the efficiency of the desulphurization process by leaking oxygen into the steel and decreasing the capability of the top slag to receive sulphur. An example of results obtained in industrial conditions is presented in Fig. 3.38 [48]. The influence of the unstable oxides is discussed more thoroughly under controlling factors. The important role played by the oxygen activity in determining the capacity of the slag to absorb sulphur has been fully discussed in Chapter 2, in conjunction with the sulphur capacity.

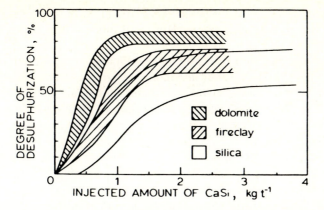

**Fig. 3.37.** Influence of ladle refractory material on desulphurization efficiency. Redrawn by [21] from [2].

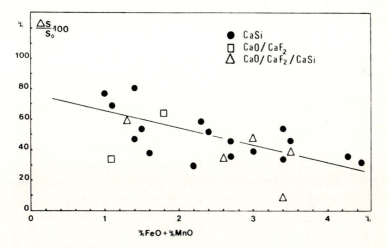

**Fig. 3.38.** Desulphurization as a function of the percent of FeO + MnO in the top slag after powder injection [48].

*The amount of top slag* has an influence on the attainable desulphurization degree as discussed. The results from industrial scale experiments reported in the literature [31, 42] indicate, for example, that under otherwise similar conditions the increase of the slag amount from 6 to 12 kg/ton yielded nearly a 100% increase in the desulphurization degree, that is, from a good 40% to nearly 80%, with the same amount of calcium injected per ton of steel.

*Controlling factors of desulphurization* are summarized here incorporating literature reviews [21, 51, 128]. All the discrepancies presented show that the transitory reactor concept [29, 30] is not properly utilized in many practical applications. In fact, it has been reported [136] that the contribution of the transitory reaction on the reaction rate of desulphurization is only 25%. The inadequate slag formation in the injection trajectory is a possible explanation for the fairly weak appearance of the transitory model and may explain the diversities in results from different plants.

But there are still other factors that make it difficult to distinguish the available reactor models [21]. The sulphur distribution depends on the sulphur capacity of the slag and oxygen activity in the steel. When aluminium-killed steel is injected with premelted slag or calcium–silicon, the above mentioned factors are very favorable for desulphurization, and they can be assumed as constant during the treatment. In contrast, these factors can be changed significantly in the top slag. First, let us examine the sulphur capacity. The top slag is made by adding lime or fluorspar on the slag-free surface, or there can be some slag left from the primary furnace. Properties of the lime and the addition practice may greatly vary from plant to plant. Generally, it can be stated that the formation of the top slag takes some time and the capability to receive sulphur is improved at the same time.

Another factor that can control the sulphur transfer into the top slag is its oxidizing power, which is defined by its iron and manganese oxide content. Considerable amounts of FeO and MnO may be present in the top slag at the start of the ladle treatment. The oxidizing effect on the slag–metal boundary can be presented by the reaction

$$(FeO) = Fe + [O]. \tag{3.9}$$

The slag is gradually reduced through an exchange reaction with the steel:

$$2\,[Al] + 3\,(FeO) = (Al_2O_3) + 3\,[Fe]. \tag{3.10}$$

The reduction is often assisted by the addition of Al powder into the slag. The change in the equilibrium sulphur distribution has been shown schematically in Fig. 3.39. Both the lime dissolution into the slag and the reduction of oxidizing components improve sulphur distribution. These reactions can take a couple of minutes. The effect of slag formation on the sulphur capacity has been presented in Fig. 3.40. The total effect of oxygen activity and sulphur capacity on the sulphur distribution is shown in Fig. 3.41. At the beginning of the treatment, the sulphur capacity of the slag can be fairly low, and the oxygen activity determined by the reaction [Eq. (3.9)]

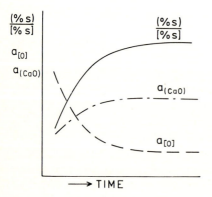

Fig. 3.39. Schematic representation of the effect of slag formation and deoxidation on sulphur distribution [51].

on the phase boundary is very high. Consequently, the sulphur distribution has a very low value, indeed. It is obvious that the ambient oxygen activity in the reaction zone is reduced quite rapidly, because experimental results show higher sulphur distributions than could be expected upon the average slag analysis [51]. With well-reduced slag and Al-deoxidized steel, sulphur ratios of several hundreds can be obtained as is shown by the arrow in Fig. 3.41. On the other hand, the aluminium oxide activity in the $CaO-Al_2O_3-SiO_2$ slags in question is far less than unity and consequently Al deoxidation can produce much lower oxygen activities than the nominal Al–O equilibrium. This has been proposed [52] as an explanation for the observed sulphur partitioning of 1000 to 3000. This has been visualised by the dashed arrow in Fig. 3.41.

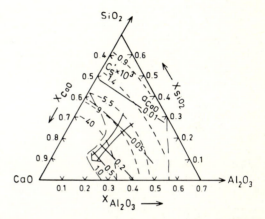

**Fig. 3.40.** Increase in sulphur capacity of slag during ladle treatment [51].

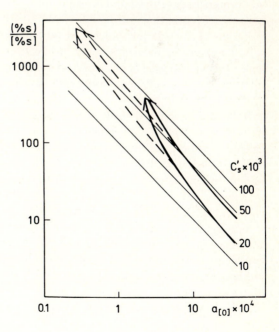

**Fig. 3.41.** Progress of the equilibrium sulphur distribution with increasing sulphur capacity $C'_s$ and decreasing oxygen activity $a_{[0]}$ [51].

What is the effect of these factors on the reactor model of injection? According to the transitoric model, sulphur transfer from steel to slag drops can proceed very far and actually very high sulphur contents have been observed in slag drops in the injection bubble zone [31]. When these sulphide-rich particles rise to the metal–slag interphase, they dissolve into the top slag. But the usable sulphur partition can be very low at the beginning of the treatment, as is shown in Figs. 3.39 and 3.41, and the slag is not capable of accepting all the sulphur from the slag drops. Resulphurization can occur according to the following reaction:

$$(CaS) + [O] = (CaO) + [S]. \tag{3.11}$$

Under these conditions, desulphurization is limited by the capability of the top slag to absorb sulphur and does not depend on the type of powder injected or on the other injection parameters.

### 3.4.2 Deoxidation

As far as deoxidation is concerned, the word actually has two meanings, namely, a decrease of both the oxygen activity and the total oxygen content. Deoxidation of the first type is a fast chemical reaction and the dissolved oxygen content may readily be lowered by deoxidants, for example, Si, Mn, and Al. Deoxidation of the second type is a rather slow physico-chemical reaction. It is strongly influenced by the flow pattern and physical properties of the system. In practice, the total oxygen content reaches equilibrium after a certain time, which is given by the ratio of the rates of removal and reoxidation.

To reach low total oxygen contents, both high removal rates and low reoxidation rates are essential. As a curiosity, it can be mentioned that the injection technology was originally developed and is also employed today for overcoming the slow rate of the reactions in the bath.

Results of deoxidation depend on very many factors, as already discussed. Dissolved oxygen, or oxygen activity in injection, has been reported in only a few investigations. *Oxygen activity* of Al-deoxidized steel has been reduced by calcium injection down to values of 5 ppm or less, corresponding to the Al–O equilibrium [2, 19, 31, 46, 67].

In Si–Mn deoxidized steels, where the soluble aluminium is very low—up to a maximum of 0.005%—oxygen activities of 10–20 ppm have been measured after calcium injection, and they are lower than Si–Mn–O equilibrium [54]. A possible explanation is that before injection treatment manganese silicate inclusions are present, which determine the ambient oxygen activity [54]. Calcium reduces the manganese out of these inclusions and modifies them to CaO–$Al_2O_3$–$SiO_2$ inclusions, which have much lower equilibrium oxygen activity.

*Slag injection* of Al-deoxidized steel has resulted in low oxygen activity compared with calcium injection [19, 40, 52, 65]. An explanation [68] to this may be that the complete deoxidation of steel with aluminium is facilitated by the simultaneous presence of molten calcium aluminate as the deoxidation product, according to the following reaction:

$$CaO + 2[Al] + 3[O] = CaO \cdot Al_2O_3. \tag{3.12}$$

For example, the steel deoxidized with aluminium alone to a level of 4-ppm oxygen activity will contain in equilibrium 0.02% soluble aluminium. On the other hand, in an aluminium and lime deoxidized melt with lime saturated aluminate as the deoxidation product, the same oxygen activity is achieved with only 0.001% Al left in solution. Thus, it appears that a $CaO + CaF_2$ injection maintains an effective deoxidation of the steel associated with a simultaneous low soluble aluminium content, a claim which at first sight appears to contain a contradiction but is in fact well supported by theoretical and practical evidence [40]. Slag injection has also been seen to intensify silicon deoxidation [69]. This effect is based on the very low activity of $SiO_2$ in $CaO–CaF_2$ or $CaO–Al_2O_3$ slags.

Oxygen activity measurement can be a very valuable tool for ladle metallurgists. Continuous oxygen activity measurements would be especially interesting in providing information about transitory effects in deoxidation, possible local deoxidation equilibria, and reoxidation processes during injection.

*Changes in total oxygen content* depend on many factors, for example, injection performance, steel grade, and reoxidation reactions. When stable ladle linings are used and primary furnace slage and atmospheric oxygen are eliminated, very low total oxygen contents, below 10 ppm, have been obtained [2, 38, 66, 70]. With less sophisticated practice, total oxygen contents of 20–50 ppm can be attained [19, 44, 45, 53, 54].

*Reoxidation* from different sources is one of the decisive factors in determining the total oxygen content of the melt. With high contents of FeO and MnO in the top slag, new deoxidation products are formed continuously in the steel during injection until the top slag is sufficiently reduced. A distinct correlation between the total oxygen content of the steel, $O_{tot}$, the composition of the top slag (FeO + MnO), and the degree of bath stirring has been established [71], see Fig. 3.42. It is clear

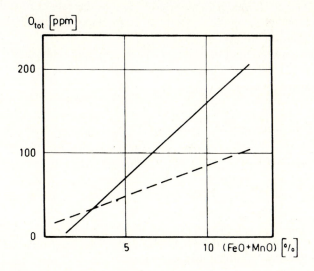

**Fig. 3.42.** The steady-state total oxygen content as a function of FeO + MnO of the top slag and flow rate of the rinsing gas. Solid line, 0.55 m³/min. Dotted line, 0.15 m³/min [71].

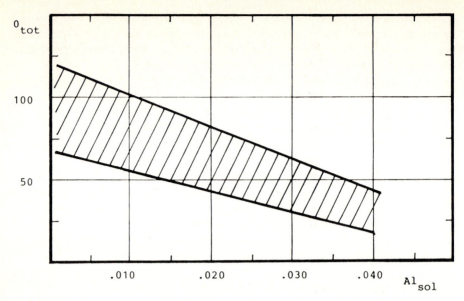

**Fig. 3.43.** Total oxygen content of the steel as a function of the content of soluble aluminium after calcium injection [5].

from Fig. 3.42 that the steady-state total oxygen content in the steel increases with an increasing FeO + MnO content in the top slag and a decreasing gas flow rate when the FeO + MnO contents are sufficiently small. An extrapolation of the curves shown in Fig. 3.42 indicates that for slags extremely low in FeO + MnO, $O_{tot}$ tends to level out corresponding to the other reoxidation sources, particularly the lining employed. It has been reported [72] that the rate of reoxidation during calcium injection can be on the order of 30–70 ppm oxygen/min in the case of 75% alumina or chamotte linings.

*Soluble aluminium* content of the steel has an apparent relationship with the $O_{tot}$ after injection treatment as shown in Fig. 3.43 [5]. This can be understood as an indirect expression of the effect of the FeO + MnO content of the top slag.

*The nature of the injected powder,* for examples, CaSi or CaO + CaF$_2$, has in some investigations been reported not to result in significantly different final total oxygen content [44, 45 66]. However, other experiments indicate [38, 40] that if the CaSi injection commences when the total oxygen content is already very low, the $O_{tot}$ will in fact increase during the CaSi injection. Further, it has been shown [38] that the $O_{tot}$ converges during a CaO + CaF$_2$ injection toward a very low final level (10 ppm or less has been obtained in cases) irrespective of the initial value. From this level, the total oxygen content subsequently starts to increase if the CaO + CaF$_2$ injection is followed directly by a CaSi injection as presented in Fig. 3.44 [40]. It has also been shown elsewhere that the final total oxygen content obtainable follows the simple formula

$$O_{tot,f} = X(O_{tot,s}) - Y(Al_{sol}) + Z(CaSi \text{ kg/ton}),\qquad(3.13)$$

where f = final and s = start.

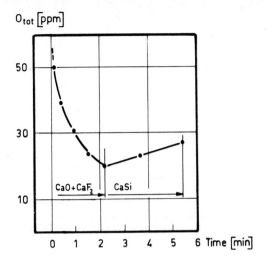

**Fig. 3.44.** Change in total oxygen content of the steel during a typical multicomponent injection of CaO + CaF$_2$ followed by CaSi [40].

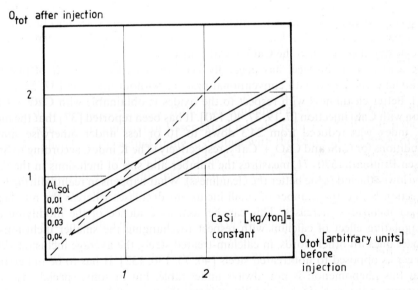

**Fig. 3.45.** The principle of the dependence of the final oxygen level on the initial oxygen and increasing soluble aluminium contents of the steel with a constant CaSi injection.

According to this expression, starting from a certain total oxygen level, the final level obtainable will be increased by injecting a larger amount of CaSi and reduced by increasing the final soluble aluminium content (see also Fig. 3.43). In practice, this indicates that the final level of total oxygen may in fact be higher than the starting level. This state of affairs is illustrated in Figs. 3.45 and 3.46. The figures show the effect of increasing the soluble aluminium level with a constant Casi

*Silicon* associated with the CaSi addition will go into solution in the steel with nearly a 100% recovery, if CaSi is used as the injected reagent. For example, if 2.7 kg/ton of CaSi is used, this would represent a silicon increase of about 0.2% in the melt. It is also possible to have silicon pickup from the reduction of the carry-over slag as well as from the ladle refractory if fireclay ladles are employed. On the other hand, Si can even decrease slightly with slag injection because of intensified silicon deoxidation. If primary furnace slag is carried over into the ladle and not removed before the injection treatment, marked changes may occur in the steel analysis.

*Manganese* will be picked up by reduction of the MnO content of the slag until below the 1% level. The increase of the manganese content of the steel may be 0.05% or even more [45].

*Chromium* content of the steel may increase as well because of the reduction of chromium oxide in the top slag.

*Phosphorus* pickup during ladle treatment is frequently reported. Phosphorus pentoxide in the top slag is readily reduced, and the P content in the steel can increase by a significant amount [43, 45, 99]. In practice, the increase of the P content can be used as a reflection of the success of the slag stopping [99].

When producing grades with very low maximum permissible phophorus content, the success in producing these grades may be dictated by the efficiency of the slag stopping. Methods related to elimination of primary furnace slag have been discussed in Section 3.2.2.1.

*Aluminium* content of the steel will be reduced during injection [43, 45, 48, 100]. This Al loss is caused primarily by the reaction of Al with unstable oxides in the top slag [101]. Aluminium loss also takes place because of reaction with the atmosphere and unstable oxides in the refractories [102].

If the slag stopping has been successful and the treatment is carried out in a dolomite ladle, Al burn-off with CaSi injection may average 0.01–0.02%. In long periods of treatment in acid ladles, the decrease in Al can be more than 0.03% [48].

A considerably large burn-off of aluminium has been reported in slag injection as opposed to CaSi use [22, 43, 45, 48, 63, 100] irrespective of the initial level of Al before injection. The burn-off may be of the order of 0.02%–0.03% or even 0.05% [22]. The reason for this very significant difference is certainly the strong deoxidizing effect of calcium in CaSi. When using CaSi at the injection, the steel and slag are supplied with an excess of deoxidizer, which effectively evens out the effects of even large differences in the initial oxygen potential. In a way, calcium protects aluminium from heavy burn-off. The rate of aluminium burn-off in slag injection can be minimized by utilizing a multihole lance for injection [103]; a threehole lance resulted in a 50% slower burn-off rate, as opposed to a single-hole lance. This difference has been attributed to the less violent conditions in the bath when multihole lances are employed.

If the injection practice is particularly good, by having little primary furnace slag in the ladle, a tight sealing lid in the ladle to avoid air oxidation, and proper refractories in the ladle, the aluminium loss during CaSi injection can be suppressed to 0.003–0.004% [4].

Burn-off of aluminium should be controlled during treatment, otherwise the final Al content may vary too much. The aluminium content can be measured by a sampling and analyzing method or by oxygen activity measurement, from which the soluble Al content can be calculated.

### 3.4.6 Behavior of Gases Due to Injection

Great attention has been paid to changes in the hydrogen and nitrogen content during injection because of the fact that these gases, either separately or together, may cause major problems in the further processing of the steel.

*Hydrogen* absorption of the order of 0.2–0.7 ppm has been reported for CaSi injection [43, 45, 48, 62]. Hydrogen originates partly from the moisture of the top slag lime and partly from atmospheric moisture. Care should also be taken to ensure that the alloys added during injection are dry. Hydrogen absorption with slag injection is generally significantly higher, that is, 1.1–1.2 ppm [43, 48]. The most important contributing factor is the $H_2O$ content of the injected slag powder. However, one has to remember that normally CaSi holds no water at all. Therefore, it appears that the hydrogen absorption from the atmosphere during CaSi injection is substantially larger compared with slag injection. Thus, it might be possible to limit the hydrogen absorption due to slag injection to at least the same level as for CaSi if the water content of the lime powder mix could be lowered by the suppliers. With careful drying and use of premelted slag, it has been possible to limit hydrogen absorption to 0.2–0.5 ppm [45].

*Nitrogen* absorption has been observed to be greater with CaSi injection than with slag injection and to vary between 10 and 40 ppm [43, 48, 62, 104]. The increase in nitrogen content during slag injection is much less, generally 5–15 ppm [43, 48, 63, 130]. Nitrogen dissolves from the atmosphere and carrier gas (if nitrogen). Therefore, steel with extremely low sulphur and oxygen contents is particularly prone to nitrogen absorption. It has been shown [105] that during injection of $CaO + 10\%$ Mg powder nitrogen is absorbed slowly in the beginning, but when oxygen and sulphur contents reach low levels, the speed of absorption increases rapidly, see Fig. 3.51. During the last minutes of injection, the recovery of nitrogen from the carrier gas may approach 100%.

The fact that much more severe nitrogen absorption takes place during CaSi injection as opposed to $CaO + CaF_2$ injection, regardless of the initial nitrogen level of the steel, must be attributed to the more violent conditions in the ladle during CaSi injection. Vaporization of calcium may cause a great deal of surface turbulence with the consequent air entrapment resulting in nitrogen absorption.

The feeding rate of the injected powder together with the amount of top slag on the steel surface before injection greatly affect the nitrogen absorption rate of the steel. These factors are summarized in Fig. 3.52 [38] in which the amount of top slag is drawn against the combined factors of the nitrogen absorption rate and the feeding rate of CaSi powder.

The following two conclusions can be drawn from Fig. 3.52. First, with a constant amount of slag, the nitrogen absorption increases with the injection intensity; and second, with a constant injection intensity, the nitrogen absorption decreases with an increasing amount of slag.

Evidently both of these factors, the injection intensity and amount of top slag,

Many of the alloying elements differ widely in their properties. For instance, elements such as Al, Mn, Ni, and Si are totally soluble in steel at 1600°C, whereas Ca and Mg have a very low solubility. In addition, Ca and Mg have a high vapor pressure at steelmaking temperatures. Another important property is the density. The density of Al, C, and Si is relatively low, whereas Ce, Mo, Pb, and W have a high density. The steel melt must be well stirred to obtain a uniform mixture of alloying elements. Elements with extremely low or high densities as well as strong oxygen affinities are difficult to introduce into the bath. Lighter elements, for example, Al, may give a high metal loss because these elements tend to remain on the bath surface and react with the oxygen in the slag and the atmosphere. Conversely, heavy elements, such as Ce, contribute to the impurities in steel because their oxides cannot be easily precipitated. Elements with a high oxygen affinity may lead to oxide inclusions detrimental to steel quality when certain types of inclusions remain in the melt.

The use of sophisticated techniques, such as injection, may be motivated if precise alloying is required. Weakly oxidizable elements, such as Si, Mn, and Cr, can be injected with approximately 100% recovery [41, 108]. It may be appropriate to add that at least Mn and Cr can be alloyed with roughly the same yield by simply adding them in lump form into the ladle. With acid ladle linings, the recovery of silicon, irrespective of the manner of alloying, often exceeds the theoretical limit because of the reduction of silica of the refractories. The recovery of injected carbon decreases with an increasing carbon content of the bath. However, below about 1% of carbon, the recovery is very close to 100% [109].

Injection of aluminium powder into a slightly deoxidized steel melt (Al ~ 0.01–0.015% before injection) gives aluminium recoveries of the order of 80–90%, see Fig. 3.53 [108].

Microalloys, such as boron, titanium and niobium, are also preferably injected into the steel. Because of the strong oxygen and nitrogen affinity of boron, the melt is first deoxidized with aluminium and the nitrogen removed through titanium addition. By injecting the titanium and boron as ferrotitanium and ferroboron,

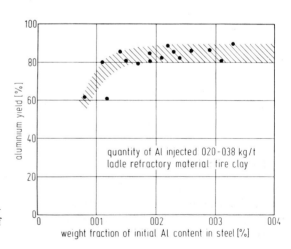

**Fig. 3.53.** Aluminium recovery as a function of the initial aluminium content of the steel [108].

yields of the order of 45–75% for titanium [41, 108] and 28–36% for soluble boron are obtainable [108]. The recovery of total boron is simultaneously 85–93% [108].

When it comes to elements such as Pb, Se, and Te, which can pose serious environmental problems, the injection process offers users an effective tool to handle the materials as well as a way to ensure high and consistent recoveries. Tests also reveal [41] that the distribution of the alloy in the end product is homogenous, a characteristic that improves material properties. Practical results have shown [108] that selenium, which is added to the steel to improve workability and especially machinability, can be injected into the steel as ferro-selenium and recoveries of the order of 71–77% can be reached.

Nitrogen is a very important element for the grain refinement of steel. The ability to form nitrides with microalloying elements, such as Al, Nb, and Ti, gives the steel improved tensile properties. Nitrogen alloying will normally be done through calcium cyanamide ($CaCN_2$) additions in conjunction with the tapping. This procedure gives a low yield (14–16%), and at the same time the analysis precision of aim will be low. The handling brings a strong dust development, causing trouble in the working environment. Another alloying manner is to use nitrated ferroalloys, for example, FeMnN.

By injection of calcium cyanamide deep (not less than 2 m) into the melt, nitrogen yields of 75–90% can be reached. The hit certainty is about $\pm \Delta N = 20$ ppm N, when the alloyed amount is on the order of 200 ppm N [110]. The contribution of the carrier gas to nitrogen alloying may be significant in certain conditions. This phenomenon has been discussed in length in section 3.4.6. Employing nitrogen as a carrier gas may yield nitrogen recoveries from the injected $CaCN_2$ of the order of 75–100% [41, 105].

## 3.4.8 Heat Losses Due to Ladle Treatment

Ladle treatment is also a useful means of ensuring strict temperature control for subsequent casting, but the temperature of the melt in the ladle progressively decreases if no external heating is applied. Therefore, the temperature losses must be known in order to determine the necessary superheating in the primary furnace. The theoretical background of the heat losses has been presented in Chapter 2.

The measured temperature drop during injection treatment in ladles of different sizes is normally of the order of 1–3°C/min (100 tons), 2–5°C/min (50 tons) and 10–15°C/min (7 tons). Apart from the scale of the equipment, which affects the surface to volume ratio, the heat losses depend on, for example, the operations performed in the ladle. Figure 3.54 gives an example of the temperature drop of the steel at OVAKO STEEL Imatra Steelworks for two different steel grades and ladle treatment practices.

The physical heat losses caused by conduction through the ladle lining and radiation from the top surface vary greatly. They depend on the preheating of the lining, its thickness and design, and the materials used. The thermal conductivity of the commonly used lining materials increases in the following order: silica, firebrick, high-alumina, and dolomite. Thus, the problems of temperature loss are greatest with dolomitic linings, and efficient preheating over 1000°C is necessary.

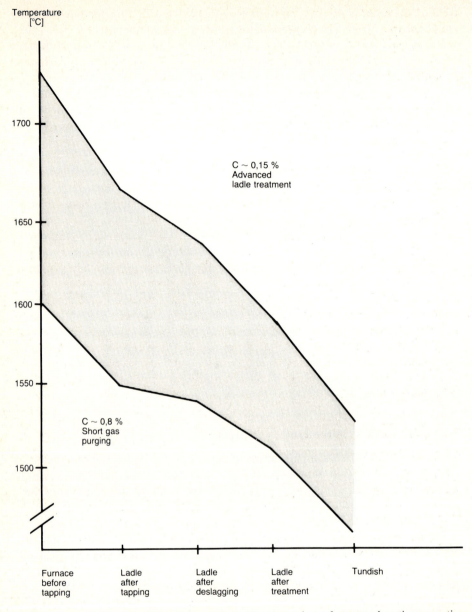

**Fig. 3.54.** Example of temperature evolution of the steel between primary furnace and continuous casting tundish at OVAKO STEEL, Imatra Steelworks.

Heat losses can further be minimized by proper lining design with a dolomitic wear-lining combined with an insulating lining [111]. Radiation losses can be reduced and made less variable by a good slag cover practice or the use of an efficient ladle cover. The lining life of dolomitic ladles is very good as compared, for instance, to acid ladles [112]. Their temperature should, as mentioned, be held constantly above 1000°C because of sensitivity to spalling if cooled appreciably between heats. Dolomite bricks also have a higher bulk density as compared to acid or alumina refractories, and this must also be taken into account.

Other heat losses are caused by the different additions made into the steel and slag phases during the course of the treatment. The cooling effect of injected powder and particularly gas is very small, only a few degrees centigrade. In addition, it can be mentioned that at the same feeding rate the temperature loss with CaSi powder is lower than that with CaO + CaF$_2$ powder. This has to be seen as a reflection of the positive heat of dissolution of elemental silicon into the steel. The heat lost because of lime and fluorspar additions to form top slag is more apparent, however. When slag amounts of 5–10 kg/ton of steel are used, the associated temperature drops are in the range of 10–20°C.

The temperature effect of alloying elements is well known [113]. A schematic presentation of temperature drops when adding different percentages of alloys into the steel is given in Fig. 3.55 [114]. Manganese and chromium have the greatest cooling effect of the ordinary elements. If only trimming additions are made into the ladle, the cooling effect of alloying may be negligible. If the need be, heat losses can be accelerated by the use of an additional coolant added to the steel ladle. This can be by the use of prepared scrap, slab dunking, or other forms of coolant, for example, prereduced pellets.

Many of these sources of heat losses are time dependent, and series of models have been developed [115] to help to quantify and isolate these sources of heat losses. In addition, the temperature from the ladle varies with time because of stratification within the ladle. It has been established [116] that thermal gradients appear in a 6-ton ladle 2 min after the disruption of argon stirring. The rate and extent of stratifications depend on the thermal state of the refractories and heat losses from the steel surface. The sources of heat losses causing stratifications are shown in Fig. 3.56 [117].

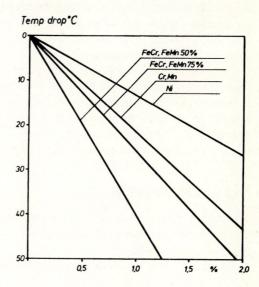

**Fig. 3.55.** Temperature drop of the steel when adding alloys in the ladle [114].

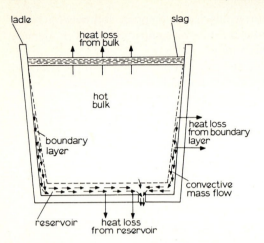

**Fig. 3.56.** Sources of thermal stratification in a ladle [117].

For this purpose, the steel in the ladle can be considered to consist of three separate regions. These are (1) a boundary layer adjacent to the walls that flows down the walls to form (2), a reservoir or stratified layer of relatively cold steel at the bottom of the ladle, and (3) a bulk of hotter steel. Heat losses from (1) are assumed to be to the ladle walls, from (2) to the bottom of the ladle, and from (3) through the slag cover only.

By consideration of the appropriate dimensional constraints and the physical properties of liquid steel [118], estimations can be made of the magnitude of these heat losses. As a result, the temperature of the steel leaving the teeming nozzle can be calculated as a fraction of time during casting, and similar calculations of steel temperature entering the mold from the tundish can be made. It must be emphasized that if the effect of stratification is ignored in the construction of temperature models serious errors are bound to occur. To avoid or, at least, minimize buildup of stratifications after the end of stirring, the use of adequate ladle covers is mandatory. To achieve a stable operation, proper ladle preheating and short circle times are helpful. Gas rinsing of the ladle during casting may be essential if the stratifications are to be eliminated completely.

Extra superheat for ladle injection, compared with normal practice with short, homogenizing gas stirring in the ladle, is not very large, It is less than the difference in ladle temperature drops between these two methods, because the temperature stability in the ladle during casting is much better after such an intensive ladle operation. For this reason, and because of good flowability of steel after Ca treatment, a somewhat lower casting temperature can be employed.

### 3.4.9 Modification of Inclusions

The modification of inclusions in the steel because of calcium treatment is widely discussed in the literature. In Japan [132], indepth surveys have been made both in basic and acid ladles.

The results of the tests regarding inclusion modification in basic ladles are shown

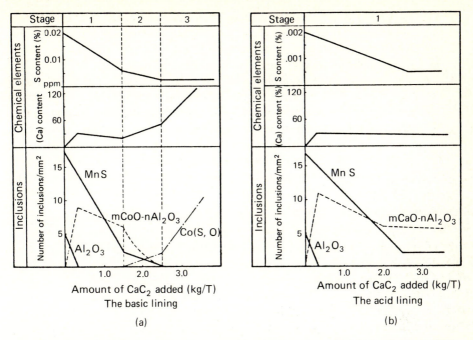

**Fig. 3.57.** Transition of elements and inclusions during calcium treatment by injection in a ladle [132].

in Fig. 3.57(a). Through these tests, it was discovered that the reactions can be divided into three stages. According to the amount of calcium added, during the calcium injection, the calcium reacts not only with the sulphur but with the oxygen and oxide in the liquid steel. The distinctive features at each stage are as follows [132].

1. *Stage one*: At this stage, not much calcium has been added. Although the desulphurization proceeds in proportion to the amount of calcium added, the sulphur content does not reach its minimum level during this stage. Consequently, one can observe MnS-type inclusions in the steel. On the other hand, the calcium reacts with the alumina, which is created by the aluminium addition before injection. Pure alumina disappears early from the liquid steel in this stage and changes into the $mCaO-nAl_2O_3$-type inclusion. The $m/n$ value increases as the amount of calcium added increases. At the final point of this stage, the $m/n$ value reaches about 2. This inclusion is called 12 $CaO$-7 $Al_2O_3$-type calcium aluminate.

2. *Stage two*: Here, desulphurization continues, only the desulphurization rate is slower than in stage one. This, of course, is because the sulphur content at the start is already low. It is not clear whether the desulphurization depends on the reaction of the calcium with the sulphur or on the removal of the calcium sulfide. At the final point of this stage, the sulphur content in the liquid steel decreases to 0.003%, and one can rarely observe MnS-type inclusions.

   The amount of 12 $CaO$-7 $Al_2O_3$-type inclusions formed during stage one diminish and $Ca(O, S)$-type inclusions are formed during this stage. In fact, at this point, the 12 $CaO$–7 $Al_2O_3$-type inclusions almost disappear, and the inclusions observed are mainly of the general $Ca(O, S)$ type. The physical size of this type of inclusions is very small.

3. *Stage three*: The changes in sulphur content are too small to be quantified during this stage. The sulphur content, however, remains under 0.003%. The calcium content in the liquid steel increases radically with the addition of calcium during this stage. Also, the number of $Ca(O, S)$ inclusions observed in steel increases in proportion to the addition of calcium. These inclusions are distributed as clusters in the final product and are composed of not only $(Ca)$, $(S)$, and $(O)$ but also of $(Mg)$ and $(Al)$, as found by inspection with an EPMA. These clusters may be caused by the reaction of the calcium dissolved in liquid steel with the brick of the ladle.

Regarding the modification in acid ladles, there is only one stage in the modification process that is similar to the occurrences in basic ladles. This has been illustrated in Fig. 3.57(b). The state of affairs is due to the lower basicity of the ladle slag. Hence, the importance of the slag basicity for calcium injection is obvious. Additional amounts of calcium beyond a certain treshold level, as indicated in Fig. 3.57(b) have no effect when acid linings are employed. In other words, there will not be any greater change in the sulphur and calcium contents or quantity of inclusions.

The best low-sulphur-containing clean steel may be obtained at the final point of stage two for the basic lining ladle. In stage one, the sulphur content is not at its minimum level and enlarged nonmetallic inclusions, such as $mCaO$–$nAl_2O_3$ are observed in the product. In stage three, although the sulphur content is always at a low level, the quantities of inclusions increase.

Considering the modification of sulphides and oxides separately in detail the following can be accounted.

*Modification of sulphides* because of calcium injection has been indisputably shown [43, 53–62]. The modifying effect of calcium has been presented schematically in Fig. 3.58 [63]. Sulphur is partly combined in detached calcium-containing inclusions, partly in oxide inclusions. Different sulphur levels have been suggested [63, 64, 130] for the upper limit, below which a complete modification of type II manganese sulphides takes place. A maximum sulphur content of 0.003–0.006% should guarantee a complete modification depending on the relative levels of calcium, sulphur, and oxygen in the steel. The amount of calcium necessary to completely suppress the manganese sulphide precipitation can be expressed with the aid of the effective calcium to sulphur ratio (ECSR) [36]. The ratio $Ca_{eff}/\underline{S}$ can determine the number of the manganese sulphide occurences. That is, when the ECSR is larger than 1.25 manganese sulphide precipitation is eliminated [36]. The effective calcium in this context is the difference between the total calcium and the calcium in oxides. It has also been reported [130] that significant modification of sulphide inclusions is achieved when the calcium to sulphur ratio is between 0.7 and 1.0. For full modification, calcium to sulphur ratios of greater than 2.0 are required, see Fig. 3.59 [130]. In this case, the sulphur content of the steel was below 0.003%, and the calcium content was the total calcium analyzed.

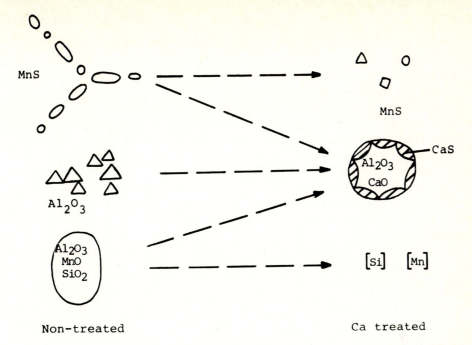

**Fig. 3.58.** Schematic presentation showing modification of inclusions due to calcium treatment [5].

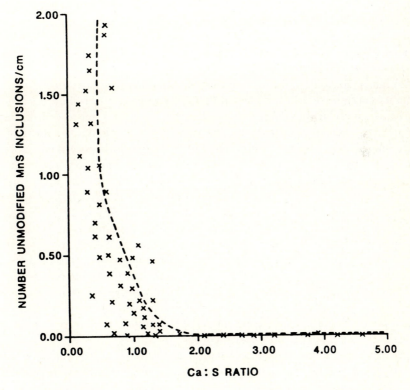

**Fig. 3.59.** Influence of calcium to sulphur ratio on sulphide modification [130].

It has also been suggested that modification of sulphides may take place even as a result of slag injection [65], but this has not been completely verified on the production scale [43, 45, 66]. Although there are no large deformable manganese sulphides in steel having low sulphur content, very small deformable, that is, unmodified, sulphides still exist.

In addition to calcium, which alters the morphology of both oxide and sulphide phases, there are elements that bring about a change of the sulphide shape only. Titanium, zircon, and rare earth metals (REM) come within this category. Again, it is important that the steel is thoroughly deoxidized and desulphurized before the addition is made. It is probably more important for the REM than for calcium treatment that basic refractories are used and that the top slag is low in FeO and MnO. Contrary to calcium, more REM remain in solution in the steel, and therefore they can react to a greater extent with the refractories, the slag and the atmosphere. All in all, the use of these elements is often difficult but can lead to improvements, particularly where an anisotropy of the material properties caused by sulphide is detrimental to the quality of the steel.

Because of the fact that the various elements mentioned act differently in influencing the sulphide shape, as well as that their effect on the mechanical properties of the steel is dissimilar, their use must be coordinated according to the required final properties of the product.

*Modification of oxides* is often desirable since, besides their amount, their type also has an important effect on steel properties. When, for example, Al deoxidation is used, alumina inclusions are formed in the steel causing, among other things, tundish nozzle clogging problems at continuous casting [82–85]. The various types of inclusions that could be present in aluminium-treated steels are shown in Fig. 3.60 [86].

Alumina inclusions, in the form of clusters, can also cause a deterioration in steel properties because of their hardness. If $Al_2O_3$ is present during rolling, the dendrites will break up giving an elongated inclusion that could be a serious surface defect. The remedy to this would be the modification of the aluminates to calcium aluminates. Modification has been observed in all investigations concerning CaSi injection. The same has been reported concerning shallow injection treatment [87]. Inclusions formed during injection of Al-killed steel contain Al, Ca, and oxygen, and often Si and S, whereas the Mn content, which is often fairly high before treatment, becomes normally very low [2, 5, 45, 53, 54, 66]. Oxide inclusions are typically surrounded by a sulphide layer, as shown schematically in Fig. 3.58.

Complex calcium aluminate inclusions are formed as molten spheres in liquid steel. They readily grow larger and separate easily from the steel. Inclusions remaining in the steel have been observed to improve flowability, and the nozzle blockage problem is avoided because liquid oxides do not have the same tendency to build up on the nozzle wall as solid alumina particles. This phenomenon will be discussed further later. Globular, discrete inclusions remain undeformed during rolling, and this can have an influence on steel properties.

Modification of alumina inclusions with slag injection has also been reported [65, 88, 130]. However, by several industrial investigations, it has been proven that the

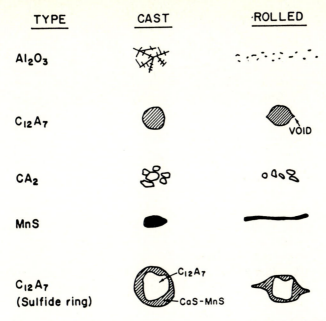

**Fig. 3.60.** Schematic diagram of inclusions in aluminium-killed steel [86].

calcium does not readily dissolve from a CaO base mix into the steel, and the modification of the aluminates thus remains nearly nonexistent [45, 66]. Flowability, however, can be improved as the absolute amount of oxidic inclusions is normally maintained at a low level because of slag injection.

It should be understood, however, that the attainable degree of modification of the inclusions depends heavily on the recovery of the calcium from the calcium carrier into the steel. A comparison of the efficiency of the various types of calcium treatments with respect to the calcium recoveries is presented in Fig. 3.61 [131]. The best calcium recoveries were obtained in this investigation with the CaSi injection and the CaFe-log immersion. The recoveries with CaFe-logs were rather scattered, however, a result that has been confirmed elsewhere [75]. The lowest recoveries were obtained [131] with injection of pure calcium granules in this comparison.

Regarding *impact on continuous casting*, calcium treatment has been shown to improve flowability of molten steel through metering tundish nozzles and to diminish casting problems caused by the tendency of the nozzles to clog [53, 54, 70, 75, 84, 87, 104, 119, 120, 121, 122]. One point is that the flowability of the steel is improved to some extent when the steel is desulphurized and deoxidized to very low S and O contents [123]. The avoidance of the nozzle clogging problem when casting, for example, aluminium-killed steels, is, however, much more important.

The nozzle clogging problem has been proven to be caused by the build up of products from strong deoxidizers on the surface of the teeming nozzle. In fact, any inclusions that are solid at casting temperatures tend to aggrevete the clogging

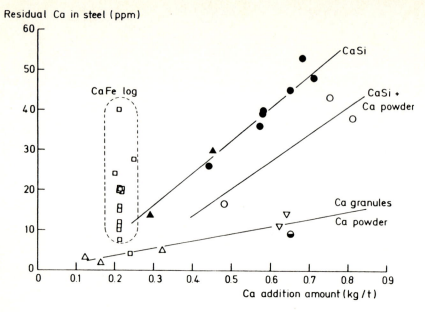

**Fig. 3.61.** Correlation between the calcium added by shown methods and the calcium recovered in steel [131].

tendency although alumina clusters are generally considered as especially harm-ful in this respect. Consequently, the Al content, for example, in 100-mm-square billet continuous casting through 13-mm tundish nozzles, is limited to roughly 0.004% [54].

The procedure of calcium treatment for attaining low-melting-point calcium aluminates is well documented in the literature [5, 38, 75, 124, 125, 132, 133]. Irrespective of this, there are still very few plants in the world actually able to cast aluminium-killed grades in routine production through metering tundish nozzles. It may be that the role of sulphur in the alumina modification has not been totally accounted.

The process of alumina modification is now described [133]. As the addition of calcium into the steel proceeds, the inclusions become increasingly rich in calcium oxide, and their liquidus decreases to the point where the calcium oxide content is roughly 48%. At some point during the treatment, the inclusions become liquid at the temperature to be used for casting. The need for a liquid inclusion sets a lower limit on the amount of calcium that must be added.

The creation of calcium aluminates affects the equilibrium of two reactions:

$$2Al + [O] \rightarrow (Al_2O_3) \tag{3.14}$$

$$(CaO) + [S] \rightarrow (CaS) + [O] \tag{3.15}$$

As the inclusions become increasingly rich in calcium oxide, a point is reached at which Eq. (3.15) becomes important. The addition of more calcium after this point

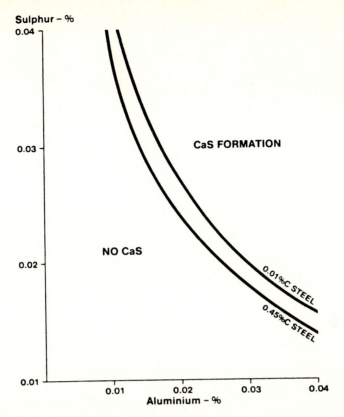

**Fig. 3.62.** The effect of Al and C contents on the critical sulphur level to avoid CaS formation at 1600°C [124].

will then create calcium sulphide rather than calcium oxide. Calcium sulphide is solid at steelmaking temperatures and has been found as a constituent of blockages in casting nozzles. Major factors influencing the possibility of CaS formation are

1. aluminium content,
2. sulphur content,
3. carbon content, and
4. temperature of the melt.

The point at which calcium sulphide starts to form sets an upper limit on the amount of calcium that can be usefully added to the steel.

Figure 3.62 [124] shows the theoretical relationship between aluminium and sulphur for two different carbon levels of the steel to give a guideline for the avoidance of CaS formation. For a given aluminium level, there is a critical sulphur content above which calcium sulphide will be formed before a sufficient amount of calcium is added to produce the optimum inclusion phase for castability, that is, the $12 \, CaO-7 \, Al_2O_3$ calcium aluminate.

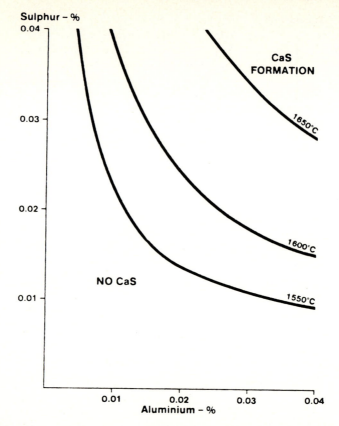

**Fig. 3.63.** The effect of temperature on CaS formation for a 0.45% C steel [124].

As the aluminium level falls, the activity of oxygen in the steel rises. This, according to thermodynamics, suppresses the formation of calcium sulphide so that higher sulphur contents can be accommodated. Figure 3.62 also illustrates the effect of carbon on this relationship, again by affecting the oxygen level of the steel.

The strong effect of the casting temperature determined by theoretical calculations is shown in Fig. 3.63. As the temperature decreases, the oxygen activity falls, and this promotes the formation of calcium sulphide.

It can be seen that if Al-killed steels are to be produced on a regular production basis, very careful control of the steel analysis and temperature as well as the calcium treatment is required. Further, the successful casting of aluminium-killed steel requires the careful prevention of steel reoxidation in the ladle and in the ladle and tundish streams. Reoxidation has been proven to increase the viscosity of the steel (because of the fact that the reoxidation inclusions are far bigger in size than inclusions of deoxidation origin) and thus to increase the tendency to nozzle clogging [82, 83, 85, 126].

## 3.5 Costs and Productivity

As far as metallurgical results are concerned and particularly the costs to obtain them, the expenses for desulphurization with CaSi injection, lime-fluorspar injection, and top slag treatment have been compared for actual production conditions [43]. The largest cost in CaSi treatment is the powder, but costs of fluxes for top slag and lance costs were also significant. In slag injection and top slag treatment, the total costs of desulphurizing agents for injection and top slag are less than in CaSi injection. Additional costs are caused by silicon alloying and markedly higher aluminium burn-off. In total, CaSi treatment was found to be the most expensive method when compared with flux injection and top slag desulphurization, the relative costs being 100:85:63.

A comparison has also been made [45] between the costs for desulphurization in the electric arc furnace and in the ladle with slag injection. The relative costs of 100 and 89 for the two, respectively, were reported.

The desulphurization costs can generally be used as a practical measure when comparing different methods. However, this is not the only criterion, since other factors in injection treatment can be quite decisive, as previously discussed. The rapid success of ladle injection can be considered as evidence of its profitability. Another reason is undoubtedly its relatively low investment cost when compared with, for example, that of vacuum equipment.

With respect to the effect of ladle injection on the productivity of the melt shop, it can be stated that steelmaking practice in the electric arc furnace, or open hearth, can be noticeably simplified when desulphurization, deoxidation, and final analysis adjustment are performed in the ladle. Furnace time can thus be shortened and the production rate increased. An example of this is shown in Figs. 3.64 and 3.65. The ladle treatment itself takes time, some 10–20 min, but this does not normally cause a delay since the operations can be suitably synchronized with the production sequence. A more detailed discussion of the economic aspects of ladle treatment is given in Chapter 4.

## 3.6 Future Aspects

The first ladle injection equipment was built at the beginning of the 1970s, and by the end of the decade, the number of such units had rapidly grown, exceeding 100 treatment stations in 1981 [128, 129].

It is most likely that ladle injection plants, possibly together with ladle furnaces, will become standard items as a unit process in secondary steelmaking metallurgy. Overall, there are many demands and challenges to develop this process further and utilize it correctly.

An important task for the future development of infection metallurgy is the better control of marginal factors in ladle treatment, for example, the strict elimination of reoxidation sources. It is by such means that inclusion cleanness problems will be

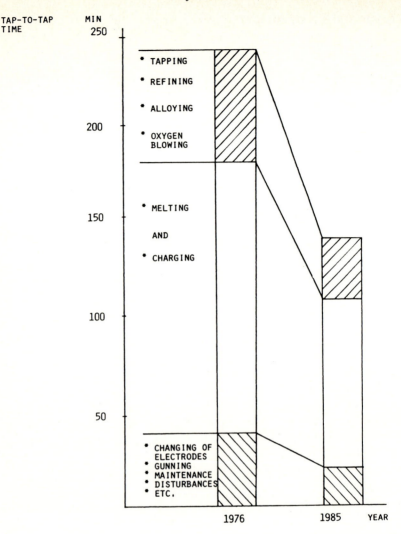

**Fig. 3.64.** Breakdown of the tap-to-tap time of the 60-ton arc furnace at OVAKO STEEL Imatra Steelworks [127].

solved. Finally, there are possibilities for process and product development that have not yet been explored. These include the general application of injection technology for the continuous casting of demanding steelgrades; new ways and means to perform the injection, for example, through the slide gate or lower sidewall of the ladle; removal of gases and tramp elements; mechanical properties versus inclusion control; improvement of cast structure; and the very promising possibility of using ladle treatment to give improved machinability, using inclusion modification and various additions and their combinations.

PRODUCTIVITY (TONS OF STEEL/MAN YEAR)

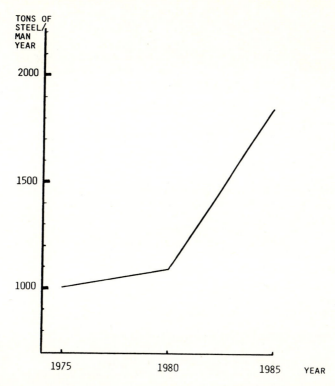

**Fig. 3.65.** Increase in productivity at OVAKO STEEL Imatra Steelworks. Commissioning of injection equipment in 1979 [127].

# References

1   T. Lehner, SCANINJECT I, Conference Proceedings, Paper 2, Luleå, Sweden (1977)
2   E. Förster, Stahl und Eisen *94* (12), 474–485 (1974)
3   K.-E. Öberg and F.J. Weiss, SCANINJECT II, Conference Proceedings, Paper 34, Luleå, Sweden (1980)
4   G. Grimfjärd, ASEA-SKE, Users Seminar (1983)
5   K. Tähtinen, R. Väinölä, and R. Sandholm, SCANINJECT II, Conference Proceedings, Paper 24, Luleå, Sweden (1980)
6   J. Bastien, Revue Métall. *74*, 325–328 (1977)
7   H. Langhammar, H. Abratis, and D. Patel, SCANINJECT I, Conference Proceedings, Paper D13, Luleå, Sweden (1977)
8   R.J. Flain, Process Engineering *53*, 88–90 (Nov. 1972)
9   F.A. Zenz, "Fluidization and Fluid-Particle Systems," Van Nostrand-Reinhold, Princeton, New Jersey (1960)
10  S. Thorildsson, Jernkontorets Annaler, *152*, 485–500 (1968)
11  S. Johansson, Report of Metallurgical Research Station-MEFOS, No. 5, 1–37 (1975)
12  Foseco International Ltd., 285 Long Acre, Nechells, Birmingham, ENGLAND B7 5JR

13  T. Lehner, Ladle Treatment of Carbon Steel, Conference Proceedings, Mc Master Symposium, Paper 7, Hamilton, Canada (1979)

14  E.F. Kurzinski, Iron and Steel Engineer 53, 59–71, (April 1976)

15  E. Schnurrenberger and P. Gerber, SCANINJECT III, Conference Proceedings, Paper 23, Luleå, Sweden (1983)

16  R.J. Fruehan, Ladle Metallurgy Principles and Practices, Iron and Steel Society of AIME Publication, 1–149 (1985)

17  Y. Iida, T. Nozaki, and F. Sudo, SCANINJECT III, Conference Proceedings, Paper 49, Luleå, Sweden (1983)

18  V. Presern, J. Arh, and T. Mlakar, SCANINJECT III, Conference Proceedings, Paper P 6, Luleå, Sweden (1983)

19  J.G. Meredith and W. Moore, SCANINJECT II, Conference Proceedings, Paper 32, Luleå, Sweden (1980)

20  A. Wikander, SCANINJECT I, Conference Proceedings, Paper 3, Luleå, Sweden (1977)

21  L.E.K. Holappa, International Metals Reviews 27, (2), 53–76 (1982)

22  T. Lehner, SCANINJECT I, Conference Proceedings, Paper 11, Luleå, Sweden (1977)

23  T.A. Engh et al., Scandinavian Journal of Metallurgy 1, 103–114 (1972)

24  K. Nakanishi et al., Tetsu-to-Hagane 64, 1323 (1978)

25  K. Narita et al., Tetsu-to-Hagane 64, S117 (1978)

26  O. Haida, K. Nakanishi, and T. Emi, SCANINJECT II, Conference Proceedings, Paper 2, Luleå, Sweden (1980)

27  K. Knop, and H-W. Rommerswinkel, Archiv für das Eisenhüttenwesen 45, 493 (1974)

28  K. Nakanishi et al., Ironmaking and Steelmaking 2, (3), 193–197 (1975)

29  H. Schenk, Stahl und Eisen 80 (21), 1377–1382 (1960)

30  H. Schenk, Stahl und Eisen 84 (6), 311–326 (1964)

31  R. Bruder, Dissertation, Clausthal, BRD (1978)

32  E. Schürmann et al. Stahl und Eisen 99 (5), 181–186 (1979)

33  J.T. Davies, "Turbulence Phenomena," Academic Press, San Diego, California (1972)

34  S. Ohguchi and D.G.C. Robertson, Ironmaking and Steelmaking 11 (5), 274–282 (1984)

35  L.W. Helle, Journal of the South African Institute of Mining and Metallurgy 81 (12), 329–337 (1981)

36  T. Usui et al., SCANINJECT II, Conference Proceedings, Paper 12, Luleå, Sweden (1980)

37  H. Kosmider et al. Stahl und Eisen 99 (22), 1215–1221 (1979)

38  L.W. Helle, B. Gabrielsson, and G. Carlsson, Scandinavian Journal of Metallurgy 15 (1), 9–15 (1986)

39  R. Källström, SCANINJECT III, Conference Proceedings, Paper 37, Luleå, Sweden (1983)

40  G. Carlsson and L.W. Helle, Scandinavian Journal of Metallurgy 14 (1), 18–24 (1985)

41  E. Spetzler and J. Wendorff, Radex Rundschau 1, 595–608 (1976)

42  P. Ritakallio, SCANINJECT I, Conference Proceedings, Paper 13, Luleå, Sweden (1977)

43  B. Tivelius and T. Sohlgren, Ladle Treatment of Carbon Steel, Conference Proceedings, Mc Master Symposium, Paper 3, Hamilton, Canada (1979)

44  K. Wada et al., SCANINJECT II, Conference, Proceedings, Paper 21, Luleå, Sweden (1980)

45  A. Moriya et al., SCANINJECT II, Conference Proceedings, Paper 16, Luleå, Sweden (1980)

46  H.P. Haastert, J. Maas, and H. Richter, SCANINJECT II, Conference Proceedings, Paper 26, Luleå, Sweden (1980)

47  J. Holt et al., Information from Paulus Lufttechnik, Federal Republic of Germany (1982)

48  G. Folmo et al., SCANINJECT II, Conference Proceedings, Paper 15, Luleå, Sweden (1980)

49  L.W. Helle, Injection Metallurgy Meeting, Instituto Argentino de Siderurgia, San Nicolas, Argentina (1984)

50  J. Otto and G. Pateisky, Stahl und Eisen 99 412–420 (1979)

51  L.E.K. Holappa, Scandinavian Journal of Metallurgy 9, 261–266 (1980)

52  H. Gruner et al., Stahl und Eisen 99, 725–737 (1979)

53  K. Tähtinen, Symposium franco-finlandais sur la siderurgia, 17–19, October 1979

54  L.E.K. Holappa and K. Tähtinen, SCANINJECT II, Conference Proceedings, Paper 17, Luleå, Sweden (1977)

55  H. Richter et al., SCANINJECT I, Conference Proceedings, Paper 15, Luleå, Sweden (1977)

56   J.P. Motte and J. Cordier, SCANINJECT I, Conference Proceedings, Paper 12, Luleå, Sweden (1977)

57   J.J. Bosley and J.J. Oravec, AIME Steelmaking Proceedings *61*, 28–35 (1978)

58   W.W. Scott and R.A. Swift, AIME Steelmaking Proceedings *61*, 38–50 (1978)

59   J.H. Mikulecky, AIME Steelmaking Proceedings *60* 338–344 (1977)

60   A. Choudhuru and G. Schmitz, Radex Rundschau, 1/2, 444–448 (1981)

61   C. Kleemann, Radex Rundschau, 1/2 449–454 (1981)

62   M. David et al., SCANINJECT II; Conference Proceedings, Paper 25, Luleå, Sweden (1980)

63   B. Tivelius and T. Sohlgren, Iron & Steelmaker *6*, 38–46 (Nov. 1979)

64   D.C. Hilty and J.W. Farrell, Iron & Steelmaker *2*, 20–27, (June 1975)

65   S.K. Saxena, SCANINJECT I, Conference Proceedings, Paper D8, Luleå, Sweden (1977)

66   M. Yoshimura and S. Yoshikawa, SCANINJECT II, Conference Proceedings, Paper 28, Luleå, Sweden (1980)

67   V. Presern. SCANINJECT II, Conference Proceedings, Paper 14, Luleå, Sweden (1980)

68   E.T. Turkdogan, Archiv für das Eisenhüttenwesen *54*, 1–10 (1983)

69   H. Sandberg, SCANINJECT I, Conference Proceedings, Paper 10, Luleå, Sweden (1977)

70   T. Ueda et al., Ladle Treatment of Carbon Steel, Conference Proceedings, Mc Master Symposium, Hamilton, Canada (1979)

71   T. Lehner et al., SCANINJECT II, Conference Proceedings, Paper 22, Luleå, Sweden (1980)

72   P. Ritakallio, Report of Swedish Ironmasters Association, B 416, 1–11 (1978)

73   W. Grimm and J. Feller, Radex Rundschau, 1/2, 455–462 (1981)

74   O. Reiber, Radex Rundschau, 1/2, 466–473 (1981)

75   R. Väinölä, P. Karvonen, and L.W. Helle, SCANINJECT IV, Conference Proceedings, Paper 23, Luleå, Sweden (1986)

76   B. Tivelius and T. Sohlgren, AIME Steelmaking Proceedings *61*, 154–171 (1978)

77   R. Scheel et al., Stahl und Eisen *105* (11), 607–615 (1985)

78   Y. Mizumo et al., XX. Internationales Feuerfest-Kolloquim, Aachen, BRD (1977)

79   W. Deilman et al., XX Internationales Feuerfest-Kolloquim, Aachen, BRD (1977)

80   H. Rellermayer et al., Stahl und Eisen *103* (10), 469–477 (1983)

81   S. Yoshino et al., Iron & Steelmaker, 16–22 (April 1980)

82   G.C. Duderstadt et. al., Journal of Metals *20*, 89–94 (1968)

83   K. Schwerdtfeger and H. Schrewe, AIME Electric Furnace Proceedings *28*, 95–102 (1970)

84   J.W. Farrell and D.C. Hilty, AIME Electric Furnace Proceedings *29*, 31–45 (1971)

85   S.N. Singh, Metallurgical Transactions *5*, 2165–2178 (1974)

86   R.J. Fruehan, Iron and Steel Society Publication, 1–149 (1985)

87   M. Devaux et al., SCANINJECT II, Conference Proceedings, Paper 31, Luleå, Sweden (1980)

88   S.K. Saxena, T.A. Engh, and H. Tveit, AIME Steelmaking Proceedings *61*, 561–573 (1978)

89   K. Kawakami et al., Tetsu-to-Hagane *69*, A-33 (1983)

90   M.F. Sidorenko, V.A. Kudrin, and N.A. Smirnov, SCANINJECT II, Conference Proceedings, Paper 3, Luleå, Sweden (1980)

91   A. Ishii et al., Iron & Steelmaker *10*, p. 35 (July 1983)

92   D. Yong et al., SCANINJECT III, Conference Proceedings, Paper 17, Luleå, Sweden (1983)

93   D. Guo et al., SCANINJECT II, Conference Proceedings, Paper 37, Luleå, Sweden (1980)

94   G. Carlsson, Report of Metallurgical Research Station-MEFOS, MF 83014 (1983)

95   H. Makar et al., Proceedings, Third Mineral Waste Utilization Symposium, Chicago, Illinois (1972)

96   R. Stone, Report, Bureau of Mines, College Park, Mdo, No. 14-09-0070-382 (1968)

97   K. Kitamura et al., Proceedings of 7th International Conference on Vacuum Metallurgy, Tokyo, 1180–1187 (1982)

98   K. Kitamura et al., Transactions of ISIJ *21*, B-469 (1981)

99   W. Pluschkell, B. Redenz, and E. Schürmann, SCANINJECT II, Conference Proceedings, Paper 10, Luleå, Sweden (1980)

100  H.-W. Rommerswinkel, Dissertation, Aachen, BRD (1973)

101  K.-H. Klein et al., Archiv für das Eisenhüttenwesen *45*, 9–16 (1974)

102  H. Abratis, Archiv für das Eisenhüttenwesen *44*, 329–336 (1973)

103  L.W. Helle and B.E. Gabrielsson, Report of Swedish Ironmasters Association, D 524, 1–32 (1984)

104   C.-E. Grip et al., SCANINJECT II, Conference Proceedings, Paper 13, Luleå, Sweden (1980)

105   R. Johansson, SCANINJECT I, Conference Proceedings, Paper 18, Luleå, Sweden (1977)

106   A.S. Kharitonov et al., Stal in English, p. 522–524, July 1970

107   S.T. Rostovtsev, Theory of metallurgical processes, Metallurgizdat (in Russian), Moscow (1956)

108   H. Abratis and H.-J. Langhammer, SCANINJECT II, Conference Proceedings, Paper 27, Luleå, Sweden (1980)

109   S. Eketorp et al., SCANINJECT I, Conference Proceedings, Paper 9, Luleå, Sweden (1977)

110   P. Ritakallio, Report of Metallurgical Research Station-MEFOS, MF 78055, 1–25 (1978)

111   D. Mehlan and E. Spetzler, SCANINJECT II, Conference Proceedings, Paper 30, Luleå, Sweden (1980)

112   K.-H. Bauer and R. Quinten, Stahl und Eisen *100* (18), 1045–1050 (1980)

113   D. Janke and W.A. Fischer, Archiv für das Eisenhüttenwesen *47*, 195–198 (1976)

114   B.Öhman and T. Lehner, SCANINJECT I, Conference Proceedings, Paper 2, Luleå, Sweden (1977)

115   W.R. Irving et al., 2nd International Iron and Steel Congress, Düsseldorf, BRD (1974)

116   J. Å. Wester, Report of Metallurgical Research Station-MEFOS, 2/68, 1–28 (1968)

117   R. Baker and W.R. Irving, Ironmaking and Steelmaking *8* (5) 216–224 (1981)

118   J. Szekely and J.H. Chan, Metallurgical Transactions *2*, 1189 (1971)

119   W.W. Scott, Jr., and R.A. Swift, AIME Steelmaking Proceedings *61* 38–50 (1978)

120   K. Nürnberg et al., Secondary Steelmaking, Conference Proceedings, London (1978)

121   K.H. Bauer Continuous Casting of Steel, Conference Proceedings, London (1977)

122   K. Öberg and F.J. Weiss, AIME Electric Furnace Proceedings *36* 54–60 (1978)

123   P.P. Arsentjev et al., Izv. VUZ Chernaya Metall. (9) 132–138 (1968)

124   I.G. Davies and P.C. Morgan, Product Improvement, Conference Proceedings, London (1985)

125   F. Faries, P.C. Gibbins, and C. Graham, Ironmaking and Steelmaking *13* (1) 26–31 (1986)

126   P.P. Arsentjev and B.S. Lisitski, Izv. VUZ Chernaya Metall. (11), 25–28 (1972)

127   H. Kalkela, K. Terho, and L.W. Helle, 2nd European Electric Steel Congress, Florence, Italy (1986)

128   L.E.K. Holappa, SCANINJECT II, Conference Proceedings, Paper 1, Luleå, Sweden (1980)

129   J. Dixmier and J. Henry, Scandinavian Lancers Club Meeting, March 1980

130   A. Herbert et al., SCANINJECT IV, Conference Proceedings, Paper 27, Luleå, Sweden (1986)

131   C. Marique, SCANINJECT IV, Conference Proceedings, Paper 25, Luleå, Sweden (1986)

132   T. Ohnishi, Secondary Steelmaking for Product Improvement, Conference Proceedings, London (1985)

133   A. Nicholson et al., Ironmaking and Steelmaking *13* (2) 53–69 (1986)

134   G.A. Irons, SCANINJECT IV, Conference Proceedings, Paper 3, Luleå, Sweden (1986)

135   M. Jehan and C. Aquirre, SCANINJECT IV, Conference Proceedings, Paper 26, Luleå, Sweden (1986)

136   Y. Hara et al., SCANINJECT IV, Conference Proceedings, Paper 18, Luleå, Sweden (1986)

138   Y. Nakamura et al., 2nd Japan–Germany Seminar, Tokyo, Iron and Steel Institute of Japan (1976)

139   H. Bode et al., Stahl und Eisen *103* (5), 211–215 (1983)

140   M. Orehoski and R. Gray, Iron and Steel Engineer *61*, pp. 40–52 (Jan. 1984)

141   K.-H. Heinen et al., Stahl und Eisen *104* (16), 49–52 (1984)

142   Injectall Ltd., Abbey House, 453 Abbey Lane, Sheffield S7 2RA, United Kingdom

143   Private communication, G. Briggs, Manager, Injectall Ltd., Abbey House, 453 Abbey Lane, Sheffield S72RA, United Kingdom

144   L. Helle and T. Persson, Report of Metallurgical Research Station-MEFOS, MF 84078, 1-42 (1984)

145   C. Bodsworth, Ironmaking and Steelmaking *12* (6) 290–292 (1985)

146   P. Ritakallio, SCANINJECT I, Conference Proceedings, Paper 13, Luleå, Sweden (1977)

# 4 Economic Considerations

## Göran Carlsson

### 4.1 Introduction

Injection metallurgy is introduced into steelmaking to reach specific objectives, such as:

1. improved steel quality,
2. increased productivity,
3. the possibility to produce special grades, and
4. increased yields.

Behind these technical objectives there are economic considerations and also strategic considerations in related markets. Local conditions in steelworks play an important role in economic calculations. Prices and interest rates are changing. Economic conditions must therefore be checked repeatedly.

### 4.2 Unit Price

The cost of different items used in the calculations in this chapter are listed in Table 4.1.

### 4.3 Consumption

A normal consumption of CaSi is 1.5–3.0 kg/tonne. If the consumption exceeds 3.0 kg/tonne, it can be regarded as a high consumption and this is usually because of the poor separation of furnace slag that is high in easily reduced oxides. In powder injection, the carrier gas consumption will not be higher than 0.10 $Nm^3$/tonne.

A monolithic lance of high alumina can be normally used for 80-min treatment. If only 5% or less of the total tonnage produced is treated, the average lance life will be considerably shorter.

### 4.4 Capital Costs

There is a trend toward more and more sophisticated equipment surrounding the basic equipment consisting of dispenser, lance, etc. As a consequence of this, investment costs show an increasing trend that is far greater than that which would be expected from inflation.

**Table 4.1.** Unit prices

| Unit | Unit price |
|---|---|
| CaSi | 100 units/kg |
| CaO/CaF$_2$ | 17 units/kg |
| Ar (internal production) | 35 units/m$^3$ |
| Ar (external production) | 110 units /m$^3$ |
| Monolithic lance | 83,000 units/lance |
| Si alloys | 50 units/kg |

The investment cost depends as much on the heat size as on the complexity of the system. A large installation can also include the automatic sampling of steel and measuring of temperature and oxygen activity, bins containing cooling scrap, and ferro alloys.

These materials are added by vibration feeders and conveyors to adjust temperature and chemical composition. From the oxygen probe analysis, the aluminium addition is calculated and then added by, for example, a wire feeding machine. The lance stand can take two lances, one for powder injection and one for gas purging. The system can be completed with a second dispenser giving possibilities for multicomponent injection.

## 4.5 Economic Interactions with Other Processes

Ladle metallurgical treatment requires extra heat to compensate the cooling effect of cold reagents, carrier gases, and temperature drop during treatment time. All heat must be generated in the primary furnace, resulting in increasing energy and refractory consumption. The extra cost in the primary furnace depends on local factors and must be determined in each case.

The ladle refractory consumption will increase with ladle metallurgical treatment. The hot, molten slag together with the stirring will result in increased wear, especially in the ladle slag line. The difference in wear may be between 10 and 30%. Alloys introduced by the reagents, such as Si in CaSi (70% Si), must be credited. An addition of 2.5 kg CaSi per tonne gives a silicon addition of 1.75 kg/tonne or 0.175%.

In many cases, ladle metallurgical treatment will increase productivity in the whole steelworks. The prerequisite for this is that the bottleneck of the whole production line is situated in the primary furnace. In other cases, productivity of the whole line will not be affected. However, in this case, time will be saved and can be used in a way beneficial for costs, such as maintenance work and lining repair.

Improved quality should manifest itself in higher reproducibility in production. General quality improvement of the products reaching the market is also beneficial but not calculable and thus cannot justify costs in production. When the yield is increased, the value difference has to be calculated between marketable products and scrap.

Ladle metallurgy is often used to produce new steel grades. The net market value is higher for these steel grades. Consequently, the net profile will also be higher.

**Table 4.2.** Operations costs[a]

| | |
|---|---|
| Variable costs | |
| Lance | 41 units/tonne |
| CaSi (2.5 kg/tonne) | 100 units/tonne |
| Argon (external production) | 4 units/tonne |
| Labor | 10 units/tonne |
| Maintenance | 4 units/tonne |
| Capital costs | |
| Investment and erection | 45 units/tonne |
| Process interaction | |
| Silicon credit | − 34 units/tonne |
| Ladle refractory | 6 units/tonne |
| Total | 176 units/tonne |
| Extra heat | 50–110 units/tonne |

[a] EAF works, 100-tonne ladle, 400,000 tonne/year, 50% injection treated.

**Table 4.3.** Benefits of powder injection

| | |
|---|---|
| Case 1: | Possibility to produce new steel grades of higher quality. |
| Case 2: | Yield improvement for scrap to solid material. It is possible to reach 4–5%. |
| Case 3: | Productivity increase because of better metallurgical optimization. An increase of 10% is possible. |

In order to summarize what has been stated in this chapter, operations costs and possible ways of utilizing the injection process are given in Tables 4.2 and 4.3.

A steelwork investing in injection equipment can usually utilize parts of all three cases given in Table 4.3. This gives a saving that is well over the operation costs.

# 5 Testing Technique for Powder Injection

## Göran Carlsson

## 5.1 Dry Blowing Tests

When powder is injected into the ladle, it is of utmost importance that the powder and gas flow rates are stable, otherwise the lance nozzles could be blocked. In order to learn how to match the properties of powder with the injection equipment, special dry blowing tests can be performed. An experimental setup is shown in Fig. 5.1.

During dry blowing tests, the powder is injected from one dispenser into a receiver. To simulate practice, the dry blowing can be performed against pressure, thus simulating the ferrostatic pressure of the steel in the ladle at the lance tip.

During the trials the following parameters should be recorded continuously:

$\left. \begin{array}{l} \dot{V}_{disp} \\ \dot{V}_{fluid} \\ \dot{V}_{ej} \end{array} \right\}$ Gas flow rate for dispenser, fluidization, and pneumatic transport, respectively

$\left. \begin{array}{l} P_{disp} \\ P_{fluid} \\ P_{ej} \end{array} \right\}$ Pressure of gas going into dispenser, fluidization, and ejector, respectively

$m_{disp}$    Weight of dispenser
$P_{hose}$    pressure in hose
$P_{nozzle}$   pressure before nozzle

The following parameters should be noted during each experiment:

length and diameter of hose,
length and diameter of lance (steel tube),
length and diameter of nozzle,
amount of nozzles,
material of hose,
pressure in receiver, and
material transported.

It has been proven that, in particular, the registration of the pressure in the transportation tube is useful for classification of the transportation. Figures 5.2 and 5.3 show, respectively, unacceptable and acceptable dry blowing results. Pressure registration is of value during ordinary injections. Unforeseen events, for example, nozzle blockage and lance failure, can be registered.

Each dry blowing should be 1–2 min. During this time, a stable pneumatic transport must occur, otherwise the powder is not suited for pneumatic transport or the pneumatic system is not built up in a correct way.

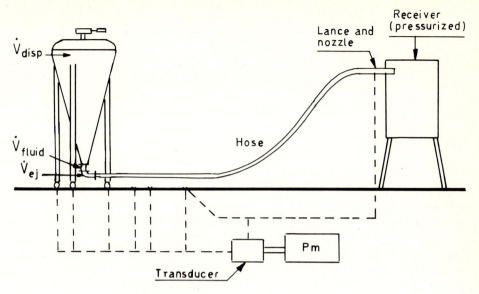

**Fig. 5.1.** Experimental setup for dry blowing tests.

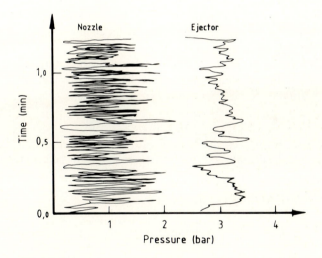

**Fig. 5.2.** Pressure curves showing unacceptably large variations.

## 5.2 Screen Analysis

For screen analysis, standardized test screens are needed. The screen consists of a circular metal frame with a seam around it so that several can be fitted on top of each other. A mesh is stretched over the metal frame and the mesh size gives the upper limit for the grains that can flow through it. The screens are placed on top

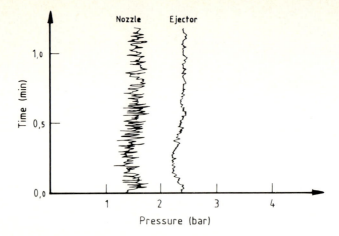

**Fig. 5.3.** Pressure curves during an acceptable pneumatic transport.

of each other on a shaker. The screen with the smallest mesh size is placed at the bottom, and the one with the largest mesh size at the top. For collecting the finest fraction that is passing through the lowest screen, a collector plate is placed under the screens.

The screen lawns consist of square holes. The length of any of the sides of a square is called the mesh size and can be given in mm. After 20 mins of shaking, the material left on each screen and on the collector plate is weighed.

Assume that $G_1, G_2, G_3 \cdots G_n$ are the weights of the material left on each screen and $G_{n+1}$ is the weight of material on the collector plate. We will then have the following equation:

$$G_1 + G_2 + \cdots + G_n + G_{n+1} = \sum_{i=1}^{n+1} G_i = G.$$

The share of each screen fraction of the total weight in percentage is calculated as follows:

$$G_1/G = f_1 \qquad G_2/G = f_2 \qquad G_{n+1}/G = f_{n+1}$$

$$f_1 + f_2 + \cdots + f_{n+1} = \sum_{i=1}^{n+1} f_i = 100\%$$

In the case of a large grain size variation, it can be difficult to obtain representative results, as these powders have a great tendency for stratification. In such a case, one has to use a "material divider." With this method, larger samples are divided into two until one comes down to a suitable amount of representative sample. Figure 5.4 shows a material divider.

The results of a screen test can be reported in two ways:

1. differential screen analysis, and
2. accumulated screen analysis.

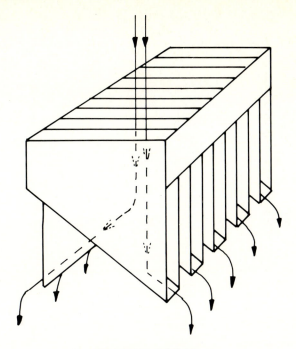

**Fig. 5.4.** Material divider.

These two types of diagrams are shown in Fig. 5.5. The time needed for one test is about 40 mins, excluding time for evaluation.

A list of different particle sizes is given in the Table 5.1.


## 5.3 Hand Test of Powder Material

How suitable a powder is for pneumatic transport can be tested, roughly but simply, with a hand test (snowball test). If the powder material is squeezed in the hand and stays together afterward, it is not suitable for pneumatic transport. On the other hand, if the powder still is free flowing after it has been squeezed, it can be pneumatically transported. The hand test is visualized in the Figs. 5.6 and 5.7.


## 5.4 Water Model Tests

In order to increase the understanding of physical phenomena involved in ladle metallurgy, water model tests can be carried out. The advantage of this kind of test is that a large number of trials can be performed in a short time, giving the observer quite a lot of informative data. The major drawback is that when modeling only a few of the modeling criteria can be fulfilled at the same time.

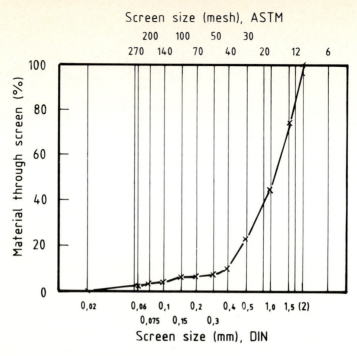

**Fig. 5.5(a).** Differential screen analysis, Al powder; note that ASTM stands for the American Society for Testing Materials, and DIN stands for Deutsche Industri-Normen.

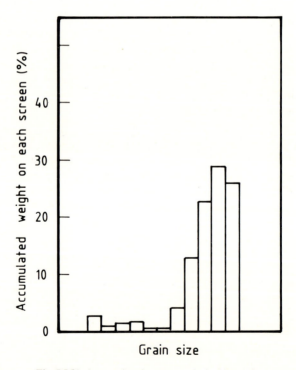

**Fig. 5.5(b).** Accumulated screen analysis, Al powder.

**Table 5.1.** Particle size in mm or mesh

| mm | British standard, BS 410/1962, mesh/inch | US standard, ASTM E 11-61, mesh/inch | Tyler, mesh/inch |
|---|---|---|---|
| 0.037 |  | 400 | 400 |
| 0.040 |  |  |  |
| 0.044 |  | 325 | 325 |
| 0.045 | 350 |  |  |
| 0.050 |  |  |  |
| 0.053 | 300 | 270 | 270 |
| 0.063 | 240 | 230 | 250 |
| 0.074 |  | 200 | 200 |
| 0.075 | 200 |  |  |
| 0.080 |  |  |  |
| 0.088 |  | 170 | 170 |
| 0.090 | 170 |  |  |
| 0.100 |  |  |  |
| 0.105 | 150 | 140 | 150 |
| 0.125 | 120 | 120 | 115 |
| 0.149 |  | 100 | 100 |
| 0.150 | 100 |  |  |
| 0.160 |  |  |  |
| 0.177 |  | 80 | 80 |
| 0.180 | 85 |  |  |
| 0.200 |  |  |  |
| 0.210 | 72 | 70 | 65 |
| 0.250 | 60 | 60 | 60 |
| 0.297 |  | 50 | 48 |
| 0.300 | 52 |  |  |
| 0.315 |  |  |  |
| 0.354 |  | 45 | 42 |
| 0.355 | 44 |  |  |
| 0.400 |  |  |  |
| 0.420 | 36 | 40 | 35 |
| 0.500 | 30 | 35 | 32 |
| 0.595 |  | 30 | 28 |
| 0.600 | 25 |  |  |
| 0.630 |  |  |  |
| 0.707 |  | 25 | 24 |
| 0.710 | 22 |  |  |
| 0.800 |  |  |  |
| 0.841 |  | 20 | 20 |
| 1.00 | 16 | 18 | 16 |
| 1.19 |  | 16 | 14 |
| 1.20 | 14 |  |  |
| 1.25 |  |  |  |
| 1.41 |  | 14 | 12 |
| 1.60 |  |  |  |
| 1.68 | 10 | 12 | 10 |
| 2.00 | 8 | 10 | 9 |

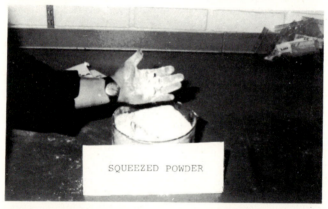

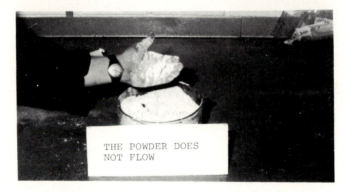

**Fig. 5.6.** Lime powder not suitable for pneumatic transport.

The thoery of water modeling has been discussed in Chapter 2. Here, a list of different tools that could be used during water modeling is given.

Velocity           Hot film anemometer
                   Strain gauge system
                   Laser

**Fig. 5.7.** Lime powder suitable for pneumatic transport.

| | |
|---|---|
| Turbulence | Hot film anemometer |
| | Laser |
| Flow pattern | High-speed camera |
| | Plastic particles |
| | Color |
| Bubble growth | High-speed camera |
| Pressure | Pressure meter |
| Mixing time | Conductivity meter |

Mass transfer     Ice rods
                  Boron oxide briquettes

Slag phase        Oil


## 5.5 Angle of Repose

A good qualitative indication of the ease of initiating flow is provided by the angle
of repose [1–6] formed by a powder heap. The angle of repose, in general, can be
defined and measured in several different ways, and the method used influences the
result. It is usually considered that powders with an angle less than 40° flow easily,
whereas those with angles exceeding 50° may form aggregates and can flow only
with difficulty, at 90° rat-holing or funneling take place.

The loose angle of repose for bulk materials, generally stockpiled, is the angle
between a horizontal line and the sloping line from the top of the pile to the base.
The angle of repose for a given material may vary, however, depending upon how
the pile is created and the density, particle shape and size, and moisture of the
material. In actual use, a stockpile will have the larger lumps tailing out at the
bottom edges of the pile and the fines concentrated toward the center. The impact
of the falling material at the top center of the pile might compact the material more
and create a somewhat steeper cone at this point.

Other angles of repose that may be found are the compacted angle of repose due
to the mechanical compaction and impact of dropping from a feed point and
consolidation due to the height of the material above in the pile and its own density.
The reclaimed angle of repose, or withdrawal angle, is that formed when material
is withdrawn from an existing pile by means of a feeder or a gate opening underneath
the central portion of the pile, generally resulting in a steeper angle because of the
amount of consolidation present in the material at the point of withdrawal. Different
angles of repose are formed by the way in which they are created. Materials
discharging from a vibrating feeder and a rotary table feeder may have different
angles at the edge of the pan or table. A rock box chute will have material piled on

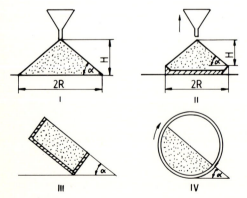

Fig. 5.8. Four main methods of measuring the
angle of repose.

each ledge, usually under some impact, which may have still a different angle of repose.

In the method of the fixed funnel and free-standing cone (I in Fig. 5.8), a funnel with the end of the stem cut perpendicular to the axis of symmetry is secured with its tip at a given height, $H$, above graph paper placed on a flat horizontal surface. Powder is carefully poured through the funnel until the apex of the conical pile formed so just reaches the tip of the funnel. The mean diameter, $2R$, of the base of the powder cone is measured, and the tangent of the angle of repose is given by $\tan X = H/R$. The method is also called the poured angle of repose.

In the method of the fixed bed cone, the diameter of the base is fixed, using a circular disk with sharp edges or a suitable machined container. Powder is poured onto the center from the funnel, which can be raised vertically until a maximum cone height, $H$, is obtained. This method is also called the drained angle of repose.

In the method of the tilting box, a rectangular box is filled with powder and tipped until the contents begin to slide (Fig. 5.8). Very similar to this method is the method of the revolving cylinder, where a sealed, hollow cylinder with one end transparent is made to revolve horizontally. It is half-filled with the powder so that the free surface of the powder forms a diametrical plane. The maximum angle that this plane makes with the horizontal on rotation of the container is taken as the angle of repose.

If a funnel is used, it has to be around 4 to 6 times the maximum particle size in diameter but not less than 10 mm. The size of the test sample is to be such that the bottom of the nozzle is 6 times the maximum particle size but not less than 100 mm above the horizontal surface when the pile is completed.

When measuring the angle of repose, five tests should be run with five different samples of the same materials and results averaged.

# References

1   A. Taskinen et al., "Powder injection properties and conditioning agents," to be published.
2   R.L. Brown, "Flow properties," S.C.I. Monograph, No. 14, 150–166 (1961)
3   D. Train, "Some aspects of the property of angle of repose of powders," J. Pharm. Pharmacol 10, 177F–135T (1958)
4   S. Frydman, "The angle of repose of Pot ash Pellets," Powder Technol. 10, 9–12 (1974)
5   B. Devise et al., "Mise an point d'une technique d'étude simplifiée de l'écoulement des puders destinées à la compression," Pharm. Acta, Helv. 50 (12), 432–446 (1975)
6   "Classification and definitions of bulk materials," CEMA Engineering Conference, Conveyor Equipment Manufactures Association, Washington D.C. 20005

# Subject Index